Dinosaurs of the Southwest

Dinosaurs of the Southwest

Ronald Paul Ratkevich

with illustrations by John C. McLoughlin

UNIVERSITY OF NEW MEXICO PRESS

Albuquerque

Library of Congress Cataloging in Publication Data

Ratkevich, Ronald Paul, 1948–
    Dinosaurs of the Southwest.

    Bibliography: p. 111
    Includes index.
    1. Dinosauria. 2. Paleontology—Mesozoic.
3. Paleontology—Southwestern States.   I. Mc-
Loughlin, John C.   II. Title.
QE862.D5R34   568'.19   75-40837
ISBN 0-8263-0405-2
ISBN 0-8263-0406-0 pbk.

FOR MY MOTHER AND MY FATHER
and
FOR MY WIFE SANDRA GRISHAM

without whose encouragement and assistance
this book would have never been written

Foreword by the Illustrator

Many readers may notice the difference between some of the restorations in this book and those to which they are accustomed. The pack-hunting *Antrodemus*, the squabbling *Coelophysis*, the hairy *Pteranodon*, and others represent syntheses of recent advances in dinosaur paleontology which allow the dinosaurs more metabolic and behavioral complexity. We have attempted to incorporate some of these newer facets of dinosaur paleontology into the text and illustration of this book; facets which indicate adaptability befitting animals which dominated life on earth for a hundred and sixty million years.

Fossil remains examined in the light of recent histological and ecological understanding point to high activity levels for many dinosaurs. Certainly, the build of many of these animals suggests a capacity for great speed and agility which does not seem consonant with a "lizard" metabolism. Furthermore, warm-blooded mammals coexisted with dinosaurs for a hundred million years and more, but the advantage of warm-bloodedness did not gain the mammals dominance until *after* the dinosaur extinctions at the end of the Cretaceous. This would seem to indicate some archosaurian metabolic advantage certainly absent in cold-blooded reptiles. Rather than portray the dinosaurs as maladapted candidates for extinction, we have chosen to restore them as animals supremely able to weather and adapt to the changes of a hundred and sixty million years. We have looked to their surviving relatives, birds, for behavioral suggestions, and have found and emphasized ecologic parallels between modern mammals and the archosaurians.

Given our comparatively short million-year stay on earth, we humans might well view with humility the span of the archo-saurians. They ruled the planet more than twice as long as have the mammals, but produced no animal with the technological capability to produce widespread extinctions. We humans have that capability and whether mammalia will achieve the lengthy dominion of the archosaurians whose history follows is question-able.

John C. McLoughlin
Santa Fe

Preface

Bits and pieces of information—much of it hidden away in turn-of-the-century scientific journals—have been published about dinosaurs in the Southwest. Many species of dinosaurs from the area have been described in technical literature, but good graphic and photographic material is not available on all of them. I have, therefore, relied on photographs and illustrations of specimens from the same geologic age and equivalent formations to illustrate some of the species which inhabited the eight southwestern states of New Mexico, Arizona, Colorado, Utah, Texas, Oklahoma, Nevada, and California. Many of the skeletal reconstructions are from early monographs and bulletins published under the auspices of the United States Geological Survey, the American Museum of Natural History, and the Smithsonian Institution. I have found the remains of many dinosaurs in the Southwest, but because of the lack of funds needed to do proper excavations, the majority remain where they were discovered to await a more healthy economic climate.

It is unfortunate that with all the fossil wealth of New Mexico, not one of the many dinosaurs collected here remains in the state. Collecting localities in general in the Southwest have been combed by outside institutions, and the fossils shipped to all corners of the world. How many of the Southwest's children will become adults without ever having the opportunity to stand in the shadow of a giant? My goal in writing this book has been to take facts about the past and make them memorable and full of meaning to the student of prehistoric animals. At the same time I have attempted to keep them acceptable in tenor and truth to the most critical scientist.

I am greatly indebted to my many friends here in the Southwest, as well as to various institutions, who have encouraged me to write this book. In particular, I must mention: John McLoughlin, a gifted illustrator and naturalist whose talents produced all of the original restorations, and whose ideas about dinosaurs changed many of my own; the American Museum of Natural History and the Smithsonian Institution whose bulletins concerning southwestern dinosaur discoveries were critical in the production of this work; the National Park Service—especially Cecil D. Lewis, Jr., superintendent of Dinosaur National Monument, who has been most helpful; Dr. Gordon Edmund of the Royal Ontario Museum for many fine photographs; Dr. Edwin H. Colbert, curator of Vertebrate Paleontology at the Museum of Northern Arizona for both his inspiration and for photographic material; my brother Richard Ratkevich for his help with research at the American Museum of Natural History; Richard Bachand, attorney-at-law, for his expert advice concerning copyrights; Marta Sherbring, who has patiently transcribed my manuscript —several times; and the University of New Mexico Office of Contract Archaeology, for giving me time off from my duties as archaeologist to do paleontological studies in the field.

Finally, and most of all, I am indebted to my wife Sandra Grisham for her unfailing aid, advice, and company both in the field and in the preparation of this book.

Contents

Foreword   vii

Preface   ix

List of Illustrations   xiii

1   Before the Dinosaur   1

2   The Triassic Period   24

3   The Jurassic Period   32

4   The Cretaceous Period   65

5   The End of an Era   100

Appendix   104

Glossary   106

Bibliography   111

Index   114

List of Illustrations

1	*Eusthenopteron*	2
2	The Grand Canyon	3
3	Geologic Time Scale	10
4	*Pteranodon*	12
5	Mosasaur (*Tylosaurus*)	13
6	*Hesperornis*	13
7	Cycads	15
8	Ammonites	15
9	Dinosaur Pelvic Structures	17
10	Diagram of Pelvic Structures	17
11	*Ornithosuchus*	21
12	Evolution of the Archosaurs	21
13	Skull of a Triassic Thecodont	22
14	Sclera	22
15	Dinosaur Family Tree	23
16	*Coelophysis* Tracks	25
17	Paleogeographic Map of North America	26
18	*Coelophysis*	27
19	Triassic Chinle Formation	28
20	*Rutiodon*	29
21	*Placerias*	30
22	Petrified Forest	31
23	Sequence of Triassic Bearing Formations	31
24	Carnosaur Footprint	33
25	Jurassic Rock Sequence Chart	33
26	The Jurassic Seas	35
27	*Camarasaurus*	36
28	*Antrodemus*	36
29	Dinosaur National Monument	37
30	Dinosaur National Monument	38
31	Sketch of *Diplodocus* Skeleton	39
32	Sauropods	42
33	*Brachiosaurus*	44
34	*Brontosaurus* Skeleton	45
35	*Camarasaurus* Skeleton	46
36	*Camarasaurus* Skull	47
37	*Diplodocus* Skull	47
38	*Ornitholestes* Skull	49
39	Quarry Sketch	50
40	*Stegosaurus* Skeleton	51
41	*Stegosaurus* Tail	52
42	*Stegosaurus* Skull	52
43	*Stegosaurus* Skeleton	53
44	Photograph of *Stegosaurus* Skeleton	53
45	*Stegosaurus* confronting *Antrodemus*	56
46	Full-scale Reproduction of *Stegosaurus* Brain	57
47	*Antrodemus* Skeleton	58
48	*Antrodemus* Skull	58
49	*Antrodemus* Skeleton	59
50	*Antrodemus* Tooth	59
51	Allosaurs surrounding a *Diplodocus*	61
52	*Ceratosaurus* Skulls	62
53	*Ceratosaurus* Skeleton	62
54	*Ceratosaurus* eating	63
55	Mammal Tracks	66
56	Cretaceous Rock Sequence	67
57	*Gorgosaurus*	68
58	*Gorgosaurus* Skull	69
59	*Gorgosaurus* Skeleton	69
60	*Gorgosaurus* Footbone	70
61	*Phobosuchus* Skull	71
62	Paleographic Map of the Cretaceous	72
63	*Tyrannosaurus Rex*	73
64	*Tyrannosaurus* Skull	74
65	*Tyrannosaurus* Skeleton	74
66	*Tyrannosaurus* Footprint	76
67	Sauropod Footprint	76
68	*Struthiomimus*	77
69	*Struthiomimus* Claw	78
70	*Struthiomimus* Skeleton	78
71	Ankylosaurs	80
72	*Ankylosaurus* Armor	81
73	*Kritosaurus* Skull	82

74	*Kritosaurus* Skeleton	83
75	Hadrosaur Scene	83
76	*Parasaurolophus* Skull	84
77	*Lambeosaurus* Skull	84
78	*Anatosaurus* Skull	84
79	*Lambeosaurus* Skeleton	85
80	*Parasaurolophus* Skeleton	86
81	*Corythosaurus* Skull	86
82	*Corythosaurus* Skeleton	87
83	*Corythosaurus* Mummy	88
84	*Kritosaurus* Tail with Tendons	89
85	*Claosaurus* Skull	89
86	*Claosaurus* Skeleton	90
87	*Prosaurolophus* Skeleton	90

88	*Pachycephalosaurus*	91
89	*Triceratops*	94
90	*Barosaurus* Skeleton	95
91	*Triceratops* Skeleton	95
92	*Monoclonius* Skull	96
93	*Chasmosaurus* Skull	96
94	*Monoclonius* Skeleton	97
95	*Monoclonius* Skeleton	98
96	*Monoclonius* Restoration	98
97	*Triceratops* Horns	99
98	Multituberculate	101
99	Map of Southwestern Dinosaur Localities	102

1

Before the Dinosaur

Geologists estimate the earth's age at slightly more than four billion years. For more than half of those years it was a hot, restless planet, busy forming rocks, water, and a highly inhospitable atmosphere. Dark, swirling clouds composed of ammonia, carbon dioxide, and other poisonous chemicals hung directly over boiling seas. Many of these gases were slowly incorporated into the earth's hardening crust, while older rocks began to free the life-supporting oxygen once tightly trapped within their molecular bonds. Uncountable combinations and recombinations of elements took place constantly. Finally, perhaps by blind chance, nature struck upon the singular combination of events and elements which produced the basis of life. The formation of complex molecular chains called amino acids was the basic step, something which may not have been duplicated anywhere else among the infinite expanses of the universe.

Fossil life preserved in ancient strata of the southwestern United States has been traced back in geologic time for an inconceivable number of years, possibly as far as twenty million centuries. For much of that time the face of our small planet underwent change as mountains rose from ocean beds and continents vanished under seas. With sudden and unpredictable rage, the awesome power of an earthquake or a volcanic eruption reminds us of the earth's present restlessness, and hints at the ancient chaos.

Functioning life-forms soon prospered in primeval oceans

whose basins have long since been uplifted and their waters drained. In this primitive, microscopic world the first one-celled algaelike plants and animals mitotically divided, increasing their number exponentially to form an ooze, or thin green veneer, which began to cover the dark volcanic rocks of warm, sunlit tidal pools. With the passing of millions of years, competition for space and food forced evolutionary change.

Without the complicated stresses of this slow metagenesis, life would have contentedly remained as it was a billion years ago. Evolution, or biological change over time, is a continuous process in which nature attempts perfection, but never seems to attain it. Environmental change has always been one step ahead of evolution, and life has continually sought new horizons. Like all horizons, once found they have proven to be only new beginnings.

Forced into change, life-forms developed which grew shells for protection, appendages for movement, eyes to see, and, in some instances, a brain and a spinal cord with which to think. It is with this last development that our story really begins. The complex forms which came out of the sea to populate the land surfaces of the earth were the vertebrate creatures. They led the way for the rise of what was to become the most incredible cavalcade of animals ever to inhabit this planet.

In the Southwest these vertebrates existed during more than 500 million years. Its seas and lowland swamps during the Age of Reptiles, some 200 million to 70 million years ago, teemed with great creatures and small. The greatest of these were the

1 *Eusthenopteron.* Advanced Paleozoic lobefin, air-gulping fish such as these were ancestral to the amphibians and ultimately to all forms of land vertebrates. About two feet in length. Illustration by John C. McLoughlin.

2   Thomas Moran drew this picture of Grand Canyon looking west from
Toroweap Point. Originally it appeared in J. W. Powell, *Canyons of the
Colorado* (1895). Reprinted from Andrew Wallace, *The Image of Arizona*
(Albuquerque: University of New Mexico Press, 1971).

dinosaurs. Today we find the record of nature's grandest experiments preserved as fossils: bits of bone, teeth, footprints, and other remains which give us a clue to the life-forms of these reptilian enigmas.

By the close of America's last great glacial Ice Age ten thousand years ago, recently in geologic terms, much of the Southwest had become blessed with the semiarid desert climate it now enjoys. With the passing of time, the parched and sparsely vegetated soils became primary targets of erosion. Wind and water cut deeply into the older underlying sediments, and arroyos, washes, and canyons exposed fossil treasures unequaled in many regions of North America. The fossil record reads much like a history book: each page is printed with the remains of unfamiliar creatures, each chapter unfolds a story so remarkable that, through the greater part of man's intellectual existence, it remained disbelieved.

Standing beside the reconstructed skeleton of a large dinosaur is a thrill for the naturalist, but a small one compared to the personal discovery of even a single fossil bone encrusted with time-hardened sediments. With the possible exception of astronomy, the science which most excites the human imagination is paleontology, the study of fossil life. The stars symbolize all that is immense and beyond the grasp of human comprehension. Fossils reach just as far into our minds. They are the only keys we possess which can unlock the distant secrets of the dinosaur's origin and ancestry millions of years before the first being awakened into human consciousness from the sleep of an animal mind.

How Fossils Are Named

In this book the terms "like" and "similar to" are used to compare one form of reptile to another. Paleontologists base their classifications on the forms which fossil remains display, and by observing similarities and differences between related fossil organisms. Simply, they match limb bone with limb bone, skull with skull, and tooth with tooth in an attempt to find enough specific characteristics to pigeonhole the animal within a group or to relate it logically to an overall evolutionary scheme. The vertebrate animals' unique skeletal structures play a critical role in this science. As the organism's changes are observed over geologic time, each anatomical variable must be compared with analogous structures of animals in similar evolutionary stages. Any modern biological classification is based on the evolutionary relationship of the parent's offspring to the offspring's progeny, and on the common ancestry of those animals which possess similar anatomical characteristics. Patterns thus determined constitute the roots of paleontology, or the biological history of fossil life.

It is impossible to speak of ancient life-forms, or even to think clearly about them, unless they are given names. It is equally impossible to examine their relationships or their place among the incredibly vast number of complex biological phenomena, or to treat them in a universally accepted scientific manner, without arranging them formally. The science of arranging, or pigeonholing, the members of the animal kingdom is taxonomy. Taxonomy, then, is the most elementary as well as the most inclusive part of animal-life studies—the most elementary because animals cannot be discussed without it; and the most inclusive because taxonomy eventually gathers together, utilizes, summarizes, and implements all that is known about the animal kingdom.

The basis of taxonomy is classification, a process which consists of grouping things according to their characteristics, placing them in a system of categories, and giving names to each established group. In zoology (and paleontology) the categorical system was adopted from the work of Karl von Linné, an eighteenth-century Swedish botanist who recognized the need for a system of nomenclature which would be universally understood, and which would order the previously chaotic biological systems.

The oldest system of classifying animals was based on ways of life, associations between animal groups, and environmental adaptations. Such classification allowed bats to be grouped with birds, and whales with fishes.

Later, systems classified animals according to the numbers and kinds of structural characteristics that the animals had in common, and their relationship to ancestral groups. The animals were placed in a hierarchy wherein each level corresponded to certain fixed characteristics. The lower the level, or hierarchic rank, the greater the number of characteristics held in common. At the highest level is the chemical basis of simple organic life. In the lower categories are the smallest units of life, each with thousands of common characteristics, although they may share fewer generalized structures exhibited in their common ancestry.

The classification system utilizes ancient Greek and Latin words as a basis for scientific notation because their meanings have remained unchanged for thousands of years. They provide an international medium of interpretation which keeps confusion within the biological sciences to a minimum.

Any practical system of classifying a large number of animals involves a hierarchy by which small units are progressively brought together into larger groups of increasingly greater scope. The following system first appeared in Linnaeus' (Karl von Linné) tenth edition of the *Systema naturae* published in 1758. It is universally regarded as the starting point of modern biological classification, and includes Empire ("Imperium"), Kingdom ("Regnum"), Class ("Classis"), Genus, Species, and Variety ("Varietas").

The following ten categories are obligatory in the classification of an animal, although the hierarchy in common use today has been greatly expanded. Only the four categories used throughout the book are generally used in nontechnical literature.

The basic taxonomic system is as follows, and I use the domestic dog as a familiar example. Notice how the subdivisions relate to each other.

Kingdom—Animal (half of the biological world)
Phylum—Chordata (those animals with a spinal cord)
Subphylum—Vertebrata (those chordates with a bony protection around the spinal cord)
Class—Mammalia (the Vertebrates with mammalian characteristics)

Infraclass—Eutheria (those mammals which give live birth via the placenta)
Order—Carnivora (those mammals which subsist on meat)
Suborder—Fissipedia (those carnivorous mammals which live on land)
Family—Canidae (all the doglike fissipedes)
Genus—*Canis* (Dogs, wolves, and jackals)
Species—*familiaris* (the domestic, or familiar dog)

In this system, each species is given a name which is composed of two, or the last, subdivisions, as, for example, *Canis familiaris*. The first term is the generic group of the specific animal. The second term modifies the genus and tells which of the species of *Canis* is being observed. Both names allude to some apparent characteristic of the animal as do the generic dinosaur names. (The modifier is often omitted, however, in general works.) For example, *Allosaurus* means "jumping reptile," the term being derived from supposed leaping in capturing its prey. *Ornitholestes* means "bird thief" and refers to this creature's habit of eating birds.

This method is useful because it enables the student of biology or paleontology to know exactly which creature belongs to the name before him. No other could possibly fit. Considering the vast numbers of living plants and animals, plus those of past ages, random naming of organisms would result in endless confusion within the biological sciences. For this reason, scientists have established strict rules by which specimens can be named, and have formed the International Committee of Nomenclature to oversee the naming processes. The Committee's rather obscure, though extremely important, existence has enabled scientists in all countries to assign taxonomic names without fear of duplication.

Geology and Time

The major difference between biology and its geologic counterpart, paleontology, is time. In order to discuss fossils and the different geologic ages in which they lived it becomes necessary to familiarize oneself with the *geologic column* and the *geologic time scale.*

The geologic column refers to the total sequence of rocks, from the oldest to the most recently deposited, in any given area. The geologic column of the Southwest includes rocks from the earliest Precambrian period to the lastest shifting sand dunes. From geologic columns previously charted on maps, the geologist or paleontologist knows exactly where in the geologic column a particular formation fits.

The geologic time scale is composed of a sequence of time units which coincide with the rock units of the geologic column, and represent the various ages of the earth's history. The time scale, worked out through various physical dating methods, is a valuable tool when working with fossils from different geologic periods.

The largest unit of geologic time is called an era. Each era is divided into periods, and each period into epochs. All of these units represent a dramatic change in life-forms from those of the preceding unit, and have been given names which describe the characteristic stage of biologic development. For example, Paleozoic means "ancient life," a name derived from the simple or ancient stage of development of the life-forms in that era. Periods are the smaller divisions of time which constitute the eras; each period derives its name from the geographical area in which it was first studied. Thus Cambrian takes its name from the Latin word for Wales (*Cambria*), while Permian comes from the Russian Province of Perm. The periods of the Mesozoic era in which dinosaurs flourished are the Cretaceous (from the Latin word meaning chalk), the Jurassic (from the Jura Mountains of Europe), and the Triassic (from the Latin *triad,* meaning three, referring to the three Triassic rock sequences in Europe).

Not only do boundaries exist between divisions of the geologic time scale because of paleontological differences, but also because of radical changes in geological reflections of former ecosystems. There are limitations to the usefulness of this scale, which become obvious when one realizes that geologic events did not always take place simultaneously with events in other geographical areas, and that animals may have evolved in distinct ways in various places.

Life-forms existing on the boundaries between time periods often have many characteristics of animals in both units. This occasionally causes some confusion. However, for convenience, we have retained this system of time and rock classification, and must work around the inherent problems of its artificiality.

The Age of Reptiles

The "Age of Reptiles" is perhaps the best descriptive phrase for the Mesozoic era, which embraces the Triassic, Jurassic and Cretaceous periods. It was during this 145-million-year span that reptilian groups dominated every habitat the earth had to offer. The land swarmed with hostile, competing reptile forms of all sizes. Streams, lakes, and the deepest seas were alive with reptiles adapted to an amphibious or completely aquatic existence. Pterosaurs, with wingspreads of fifty feet or more, ceaselessly glided on gentle offshore air currents in search of surfacing fish. Of the fourteen separate reptilian orders that lived during the Mesozoic, only four survive today.

Evidence indicates that during most of this Age of Reptiles— with the exception of much of the Triassic period—the climate remained tropical or subtropical. The abundant fossilized plants recovered from coal-bearing strata in Arizona and New Mexico show a close affinity to plant species living today in hot, wet zones of southern Florida or Central America, where the climates approximate that of the Mesozoic. Dense swamps in these areas remain in their own Age of Reptiles. Cycads, flowering gymnosperms related to the living cycad forms, were plentiful and conspicuous. Their low palmlike trunks with highly ornamented bark and fernlike leaves dominated large areas which resembled commercial pineapple groves. True ferns carpeted the rich soils of low, wet regions. Large palm trees, scale trees, and many varieties of primitive flowering plants became abundant, and hardwood

3 Geologic Time Scale. Illustration by John C. McLoughlin.

Era	Period	Representative Animal Life
Cenozoic 70 million years	Recent	
	Quaternary 1,000,000	
	Tertiary 69,000,000	
Mesozoic 155 million years	Cretaceous 65,000,000	
	Jurassic 45,000,000	
	Triassic 45,000,000	

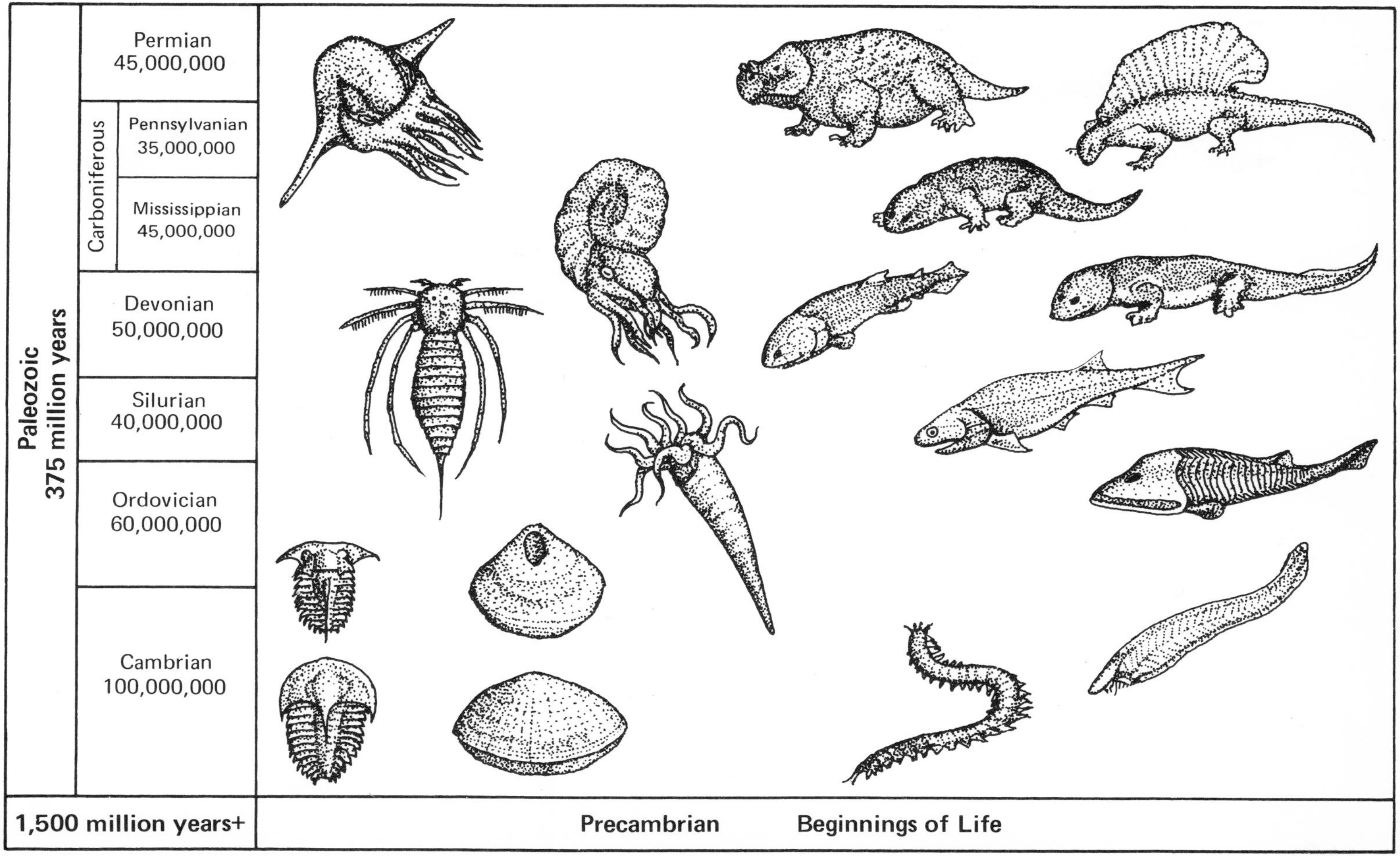

Permian 45,000,000						
Pennsylvanian 35,000,000						
Mississippian 45,000,000						
Carboniferous						
Devonian 50,000,000						
Silurian 40,000,000						
Ordovician 60,000,000						
Cambrian 100,000,000						

Paleozoic
375 million years

Precambrian   Beginnings of Life

1,500 million years+

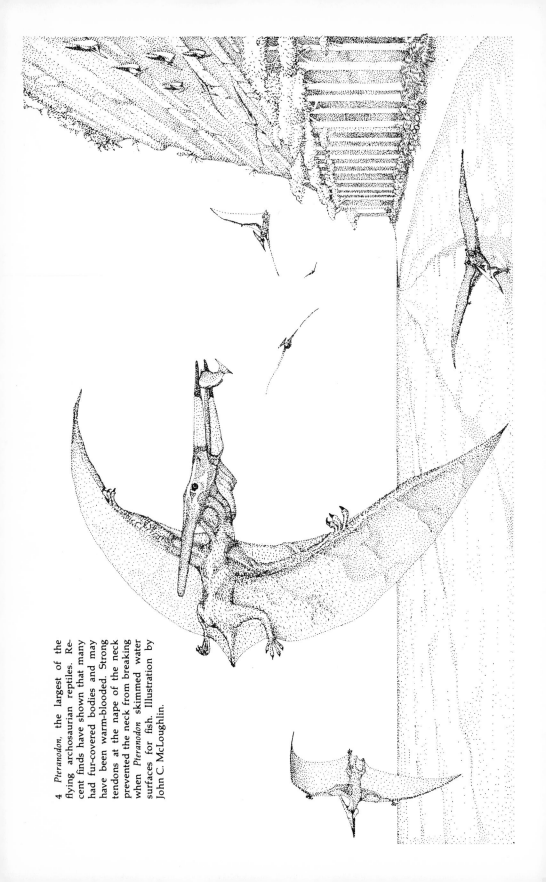

4 *Pteranodon*, the largest of the flying archosaurian reptiles. Recent finds have shown that many had fur-covered bodies and may have been warm-blooded. Strong tendons at the nape of the neck prevented the neck from breaking when *Pteranodon* skimmed water surfaces for fish. Illustration by John C. McLoughlin.

5 *Tylosaurus*, an archosaurian reptile, was not an aquatic dinosaur as popularly believed. Upper Cretaceous mosasaurs like this grew to fifty feet or more in length. Illustration by John C. McLoughlin.

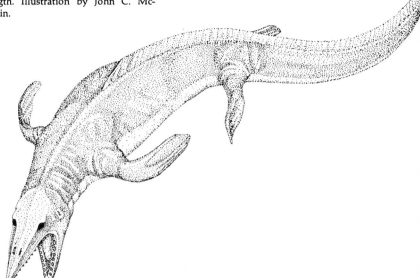

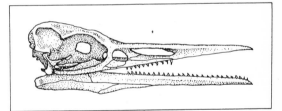

6 *Hesperornis*, a large bird with teeth, was a powerful swimmer but could not fly. About the size of a loon. Illustration by John C. McLoughlin.

forests developed and spread over many of the higher areas. Despite what is shown in many Mesozoic environmental reconstructions, only a few grasses had evolved, and none was an effective ground cover.

Oceans were inhabited not only by huge aquatic reptiles, but also by giant predaceous cephalopods, some of whose nautilus-like shells grew to be six or more feet in diameter. Certain living cephalopods—squids, for example—grew to forty or fifty feet in length and were known to attack large whales. Invertebrates of this size may have been observed by fifteenth-century sailors who subsequently reported them as sea monsters.

Volcanism was widespread during the Mesozoic. Great quantities of fly ash were shot up from the earth. Remaining in atmospheric suspension, they may have produced widespread greenhouse effects which contributed to the formation of the moist, warm climates well suited as reptilian habitats.

By the middle of the Age of Reptiles, both true mammals and birds were fully developed, having evolved from an archosaurian subclass comprised of all Mesozoic "ruling" reptiles. By the late Mesozoic, both birds and mammals were geographically widespread. And, by the end of the Cretaceous period the mammalian population had exploded. In northern New Mexico's Ojo Alamo beds the remains of quantities of primitive mammals, including the multituberculates and possibly marsupials, dating from the Cretaceous have been found.

What Is a Dinosaur?

In 1841 in an address to the British Association for the Advancement of Science, Sir Richard Owen coined the name dinosaur. He derived it from two Greek words: *deinos*, meaning terrible, and *sauros*, meaning lizard or reptile. Although many dinosaurs were the frightful giants Owen first envisioned, not all species warranted the term "thunder lizard," a graphic description based on the quaking of the earth where dinosaurs walked. Some European dinosaurs were not much larger than a barnyard

7   Conspicuous flowering gymnosperms which became dominant members of the Mesozoic ecological system. The cycad shown above provided food for one of the many herbivorous dinosaur forms. Illustration by John C. McLoughlin.

8   Cephalopods such as the ammonites shown here were active predaceous invertebrates which dominated the Mesozoic Seas. One inch to seven feet in diameter. Illustration by John C. McLoughlin.

rooster, while in America during the same period, others reached lengths of eighty to ninety feet. The word *dinosaur*, originally used to describe all known dinosaurs, is still used as a popular name for two separate biological orders of extinct reptiles, the Saurischia and the Ornithischia. Unfortunately, it has led many people to believe that all dinosaurs were part of an all inclusive group. Any close relationship between the two dinosaur orders is purely superficial, and exists only in the fact that both groups walked with all limbs structurally under their bodies. All other reptiles sprawl, with their humerus and femur bones nearly parallel to the walking surface. Each order experienced separate and distinct evolution, and the two are no more closely related than is a crocodile to a winged pterodactyl.

The saurischians derived their name from the reptilelike structure of their pelvic bones (Greek: *sauros*, lizard; *ischios*, pelvis). Most members of this order, like the well-known giant *Tyrannosaurus rex*, were ferocious carnivores (more precisely carnosaurs). A few of its members were four-footed, meat-eating, amphibious creatures, some of which became the largest animals ever to live on land.

Ornithischian dinosaurs, on the other hand, were strictly herbivorous, or plant-eating. Their name was also derived from a description of pelvic structures, which in this order was birdlike (Greek: *ornithos*, bird; *ischios*, pelvis). Out of this group evolved horned, duck-billed, armor-plated, spined, bone-headed, and beaked species. Few, however, reached the immense sizes of the lizard-hipped varieties. Many ornithischian dinosaurs were quadrupedal, with their hind legs considerably larger than their front legs, a characteristic which had been retained from their earliest ancestors, most of which presumably walked erect. The fossil record substantiates this. In fact, the ancestral path of the ornithischian leads back to an early Triassic thecodont (socket-toothed) reptile which could rise from a quadrupedal stance while walking to a bipedal gait while running.

Skulls of some of the dinosaurs were half a dozen or more feet in length. Housed within them was, in many species, a brain no larger than a walnut—with the primary function of operating the reptile's jaws while feeding. The capacity of the reptilian brain is conspicuously finite; it serves mainly to transmit sensation to other areas along the spinal column. The mammalian brain, by contrast, is quite large in relation to overall body size, and is

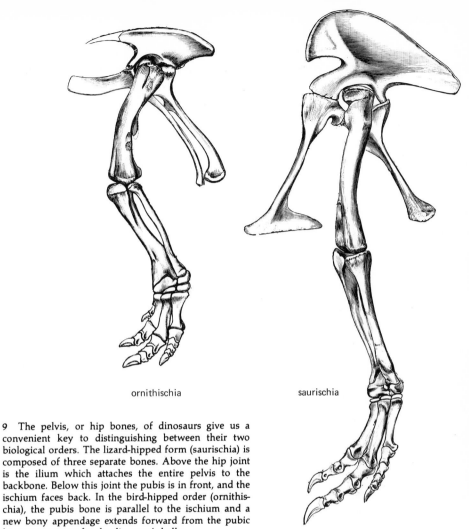

ornithischia   saurischia

9   The pelvis, or hip bones, of dinosaurs give us a
convenient key to distinguishing between their two
biological orders. The lizard-hipped form (saurischia) is
composed of three separate bones. Above the hip joint
is the ilium which attaches the entire pelvis to the
backbone. Below this joint the pubis is in front, and the
ischium faces back. In the bird-hipped order (ornithischia), the pubis bone is parallel to the ischium and a
new bony appendage extends forward from the pubic
base as a support for the dinosaur's belly.
    The term bird-hipped is used only to describe the
appearance of these bones and does not imply a
biological relationship beyond a common ancestry.

10   Diagram of pelvic structures.

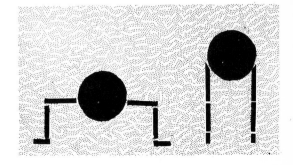

composed of a nearly infinite series of complex electrical connections which control thoughts, instincts, impulses, and reactions, and possesses the uniquely mammalian capability of interpreting messages passed on from the senses.

Were dinosaurs nothing more than highly specialized reptiles? There is growing evidence that this may not be true. At the very least, paleontologists are reevaluating their ideas concerning the metabolic makeup of the smaller, birdlike dinosaurs which held sway for 155 million years even in the presence of homothermic (warm-blooded) mammalian precursors and true mammals themselves. This dominance by dinosaurs is in itself startling. It points up the fact that for 140 million years mammals remained tiny and unspecialized. Given the chance, mammals, being adaptable creatures, will fill every environmental niche that nature offers. Could certain of the archosaurians have held such an ecological advantage over early mammals if they were cold-blooded creatures subject to temperature fluctuation?

Were some dinosaurs, then, homotherms capable of competing successfully with mammals? It must be taken into consideration that the many similarities between the structure of modern reptiles and the remains of dinosaurs naturally led the first paleontologists to conclude that these were great lizards. Take for example the archosaurian crocodile, which is a classic reptile, and the closest living relative of the dinosaurs. Is it cold-blooded? Not exactly, at least not the way a snake or turtle is. Crocodiles possess the rudiments of homothermic metabolism and a heart constructed of four chambers (as in all warm-blooded mammals and birds), and, unlike any other reptiles, are protective of their eggs.

Birds—nothing more than "glorified feathered reptiles"—are intensely parental. Moreover, although their brains are comparatively small, they are capable of a great variety of instinctual behavior which sometimes approximates "intelligence" in its complexity. The areas of the brain dealing with homothermic metabolism are small enough to have been incorporated into the archosaurian brain, and with the aid of other nerve centers similar to their brain, the small archosaurians could easily have been homotherms—albeit dim-witted and instinctually driven. Interestingly enough there is a possibility that small, active dinosaurs were feathered, as are birds. This would have helped them maintain a constant body temperature. Hard evidence, however, does not yet exist on this point.

Where then is the cutoff point between large and small dinosaur species, the point at which these reptiles lost their ability to operate with a homothermic metabolic system? Perhaps, as with the modern crocodilians, the larger dinosaur forms were partially warm-blooded. Even if giant dinosaurs lacked the true temperature control of mammals and birds, their large bulk alone offered an interesting alternative. Large dinosaurs like the sauropods required a long time to warm up and cool down; they did not react to daily fluctuations of temperature. *Brachiosaurus*, for example, had no trouble heating up its fifty tons of "cold-blooded" tissue. Its body's own physiological processes, produced by the friction of its heartbeat (the heart weighed a hundred pounds or more), by muscles rubbing against muscle and bone, and by the very breakdown of food, were sufficient for thermal control.

The giant sauropods were probably more affected by cooling off than warming up. The heat which a reptile generates is proportional to its own weight, but the rate at which it loses heat is proportional to the surface of the skin. If the length of the dinosaur were to double, the surface of the skin would be increased four times and the weight of tissue, eight times. This means that the larger the beast, the longer it takes to cool off.

Dinosaurs were more than just highly specialized reptiles. They were great, dim-witted, eating machines, whose existence has sparked the imagination of people since 1824, when English geologist William Buckland first described giant fossil reptile bones in his publication "Notice on the Megalosaurus or Great Fossil Lizard of Stonefield." The dinosaurs have aroused, and continue to arouse, more scientific conjecture than any of the modern or fossil reptiles.

Ancestor to the Dinosaurs

The ancestry of dominant Mesozoic reptile orders goes back to the closing millennia of the Paleozoic era, the Age of Fish, some 225 million years ago. Such varied groups as crocodiles, flying reptiles, and dinosaurs were evolving. All are biologically grouped within the taxonomic subclass Archosauria, or the ruling reptiles.

As seen in all the later archosaurian forms, there was a continual tendency toward a bipedal, upright posture. The dinosaurs' ancestors, the thecodonts, or socket-toothed reptiles, developed from primitive reptiles of the Pennsylvanian period, whose awkward lizardlike sprawl was slowly modified. Various skeletal elements changed, enabling those advanced reptiles to move with a semierect, two-footed gait, a major characteristic of many later archosaurs. The quadrupedal, or four-footed, dinosaur forms have obviously regressed from the upright stance inherited from their earliest progenitors. Their hind limbs remained much longer than those in front, giving them a rather humped appearance.

The entire weight of the biped's body was supported by the hips rather than by a combination of skeletal elements. Its legs did not spread out at the sides of the body but rather developed into forward-facing postlike appendages extending from the hip to the ground, without much bend at the knee.

The head of the femur, which was flat in early reptiles, developed a smooth ball joint which fit snugly into the socket, or acetabulum, of the hip. In this way, strong support was added to a skeleton designed for speed and agility. Early thecodonts, however, were only semibipedal and probably rose on their hind limbs only when the necessity for speed occurred. Several modern species of reptiles—the collared lizard of the American Southwest, for example—often take off with a rapid bipedal gait when frightened, but because of limitations in structural design, always return to all fours at the end of a sprint. None of the lizards ever became true bipeds. The ruling reptiles belonged to a different stock but their history parallels that of lizards.

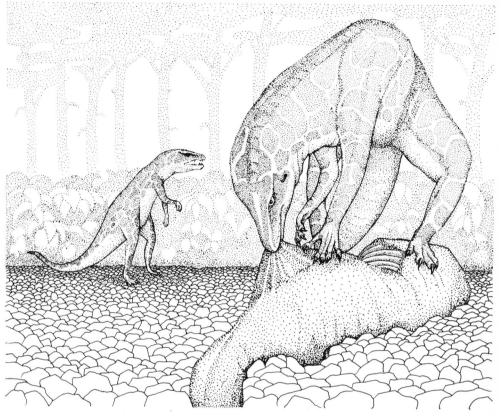

11 *Ornithosuchus*. These archosaurian reptiles—resembling large modern lizards—were sharp-toothed predators who relied on speed to catch their prey. The presence of rows of small bony plates down the length of its back shows the beginnings of armor development. About three and three-quarters feet in length. Illustration by John C. McLoughlin.

12 Evolution of the Archosaurs. Notice the abrupt termination of most forms at the beginning of the Cenozoic era. After Romer.

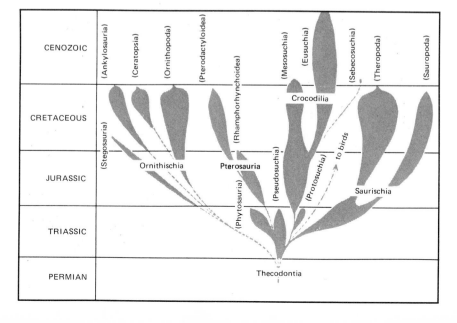

Part way down the femur, or upper legbone, the thecodonts developed a new attachment called a trochanter for connecting the strong muscles which originated at the base of its tail. The long extension of this limb muscle added additional pulling power to the hind limbs, and subsequently speed.

With the development of this new upright stance the need for strong front limbs diminished; a gradual reduction in the usefulness of the thecodont's "arms" can be seen in the evolutionary fossil record. They become short and frail. At the same time, a large mouth and pointed teeth developed to replace hands and long grasping fingers.

The thecodont's skull was generally long and slender with teeth set in sockets, as was all dentition in later ruling reptiles. The eyes were large and well developed. As a strengthening device, a ring of bony plates, or sclera, grew in a circle around the pupil, and like the diaphragm of a camera, controlled its expansion and contraction.

Thecodonts then were the common ancestors of the two great orders of dinosaurs, as well as of crocodiles, flying reptiles, and birds. All of these groups possess some of the thecodonts' characteristics, and today's surviving descendants still retain many.

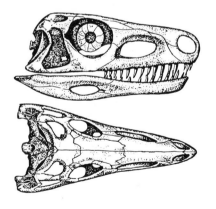

13　Skull of Triassic Thecodont. About six inches in length. From Broom.

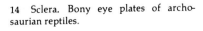

14　Sclera. Bony eye plates of archosaurian reptiles.

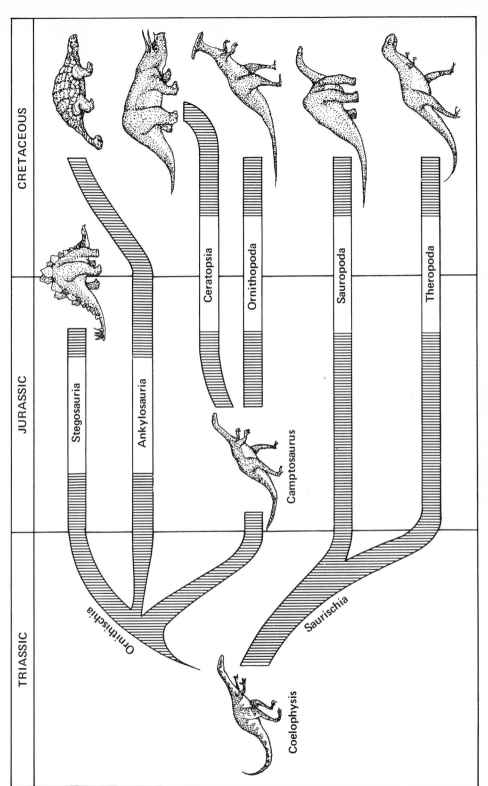

15   Dinosaur Family Tree.   Illustration by John C. McLoughlin.

2

The Triassic Period

The Little Dinosaurs of New Mexico, Colorado, and Utah

The Triassic, which began some 225 million years ago, was the first of the three geologic periods that constituted the Mesozoic era, or the Age of Reptiles. The name Mesozoic was coined by early students of historical geology who considered it to be the middle-life era (Greek: *mesos,* middle; *zoon,* life). The Mesozoic is preceded by the Paleozoic era (the Age of Invertebrates) and is followed by the Cenozoic (the Age of Mammals).

The Triassic saw the rise of most ruling reptile groups, including the earliest varieties of dinosaurs. Rocks of this age in the Southwest are easily recognized by their prominent, deep red siltstone cliff faces. Since hot desert conditions tend to leach out iron oxide, these rocks reflect the generally semiarid climate which predominated during their formation. In Arizona and Utah there are red rock beds measuring over 4,000 feet.

There are eight distinct formations of Triassic rocks in New Mexico, Arizona, Colorado, Utah, and Texas, each representing succeeding stages of geologic time and biological change, and each containing its own distinctive fossil record. Of these Triassic formations, which include the Moenkopi, Shinarump, Dockum, Chinle, Wingate, Tecovas, and Trujillo—only the Chinle sandstone and shales of Arizona and New Mexico contain actual fossil remains of early dinosaurs.

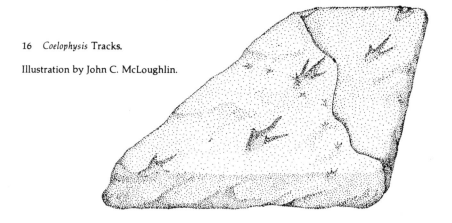

16   *Coelophysis* Tracks.

Illustration by John C. McLoughlin.

Order—Saurischia
Suborder—Theropoda
Family—Podokesauridae
Genus—*Coelophysis*

Coelophysis, the earliest of the Triassic dinosaurs, was small, weighing perhaps no more than forty or fifty pounds. Its hollow, birdlike bones added little to its overall weight. When standing erect on its strong hind legs, it was perhaps three feet tall and measured about ten feet from the tip of its well-equipped mouth to the end of its slender, whiplike tail. Its short front limbs were well adapted for grasping prey—the smaller lizards and amphibians on which it fed. *Coelophysis* was apparently also cannibalistic, for within the rib cages of some of these fossil skeletons were the remains of tiny, immature *Coelophysis* which had been swallowed, in reptile fashion, whole.

The first fossils of these little dinosaurs were collected near Abiquiu, New Mexico, during the middle 1880s by a professional fossil collector working for the Philadelphia Academy of Natural Sciences. David Baldwin's discovery, which was later named *Coelophysis,* was fragmentary. It was not until 1947, when George Whitaker of the American Museum of Natural History explored the same area, that new, more complete specimens came to light. His was no small discovery, for within a thin, two-foot layer of

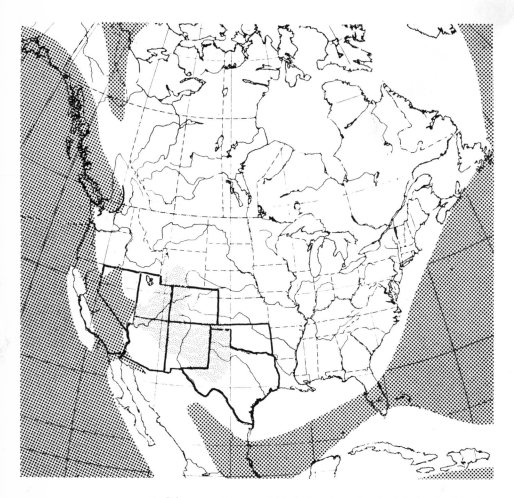

17 Paleogeographic map of North America as it existed during the Triassic
period. During the Late Triassic, most of the Southwest had a semiarid
climate similar to today's. Local areas were, however, well watered and
apparently subtropical or temperate. The above map indicates the extent of
the Triassic Seas (shaded) and areas of continental sediment deposition
(stippled). It is in these land sediments that the earliest dinosaur remains
have been found.

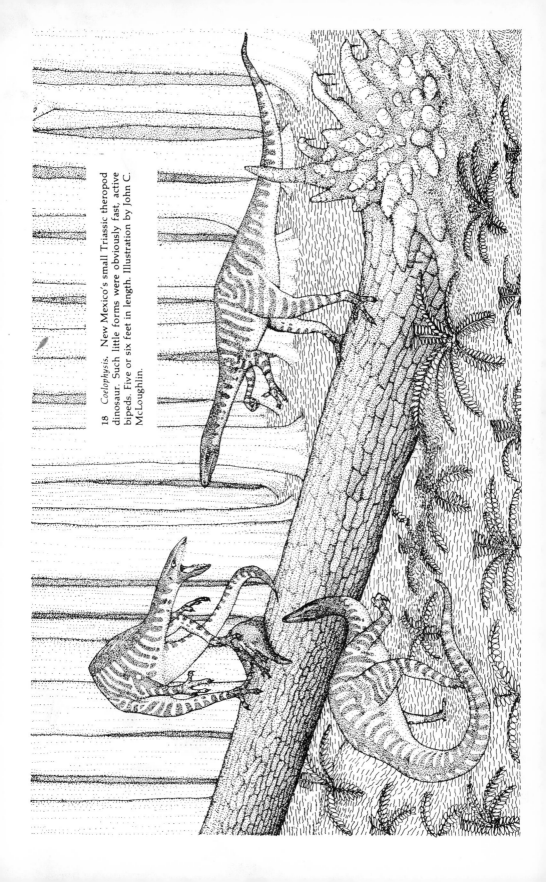

18 *Coelophysis.* New Mexico's small Triassic theropod dinosaur. Such little forms were obviously fast, active bipeds. Five or six feet in length. Illustration by John C. McLoughlin.

19 Triassic Chinle Formation, Ghost Ranch, New Mexico, in the low hills. The cliff is Jurassic Entrada Sandstone and Todilto Formation, and above it is Jurassic Morrison Formation. On the skyline is Cretaceous Dakota Formation.

stone, Whitaker found the skeltons of "dozens and possibly hundreds" of dinosaurs "piled one on top of the other and crisscrossed in every direction."

How so many dinosaurs of the same species died and fossilized in this small area is an interesting question. One could imagine that this deposit centered around what might have been the last water hole at the end of an unrelenting dry season. *Coelophysis*, which surely had the acute senses of modern reptiles, "smelled" this moisture and may have traveled many miles to find what unfortunately was nothing more than a damp mud wallow or a highly poisonous alkali pond. One by one the little dinosaurs perished, falling into a muddy tomb which would hold their bones for millions of years.

The environment in which *Coelophysis* lived must have been rather stark; fewer than 400 species of land plants have been found in Triassic sediments. It would not be unreasonable to assume that harsh desert conditions of the time could not support lush vegetation except close to local water sources. Those plants which did exist on the desert sands were, in most instances, rapidly desiccated and blown into dust by hot desert winds. There exist, however, several deposits within the western red beds where abundant plants have been preserved. The Petrified Forest of eastern Arizona has within its Chinle shale the agatized logs and stumps of giant conifers, pines which in some instances stood over 200 feet in height when alive. Close to the ground were true ferns and cycads, which held the ecologic niche of the yet to exist grasses. Towering tree ferns, which developed a thick, woody trunk and reached heights of forty or fifty feet, were holdovers from the preceding Paleozoic era. Scouring rushes—which gave cover to amphibians and aquatic reptiles—were common weeds along waterways.

Curious crocodilian reptiles (not closely related to dinosaurs or crocodiles) frequented streams during the southwestern Triassic. These were the phytosaurs. The phytosaur's skull was narrow and greatly elongated. The jaws of the typical example, *Rutiodon*, were armed with hundreds of sharp teeth, well adapted for catching and eating fish. Two nostrils were located high on the skull between the eyes, an arrangement which enabled the phytosaur to remain submerged while searching for fish. One species from western Texas grew to a length of twenty-five feet.

Mammallike reptiles were also common during the Triassic. The remains of one species, *Placerias*, were found in abundance near St. Johns, Arizona. Near the base of the Chinle formation there, nearly sixteen hundred of its bones (all representing dismembered carcasses) were discovered in one thin layer of shale. In South Wales, geologic formations of roughly the same age contained the first true mammals, which evolved from mammallike reptiles.

20 *Rutiodon.* Archosaur, much like a crocodile, that was a common reptile of the Southwest's Triassic waterways. Illustration by John C. McLoughlin.

21 *Placerias*. Reptile that looked like a mammal, from the Chinle Formation of St. John's, Arizona. Illustration by John C. McLoughlin.

By the end of the Triassic period, the archosaurs, or ruling reptiles had become the dominant land vertebrates. Thecodonts such as the *Ornithosuchus* gave rise to both dinosaur orders but became extinct themselves by the period's close. Ornithischians, the bird-hipped dinosaurs, were slow to evolve during this time, and the order is only represented by rare fragmentary specimens. Saurischians, however, became abundant in the Southwest and forms within the order, like *Anchisaurus* and *Plateosaurus*, represented the early ancestors of the giant sauropods which were to follow in the Jurassic.

22 Petrified Forest National Park, Arizona. *Araucarioxylon,* an extinct conifer, accounts for nearly ninety percent of the petrified wood found in the Park's Chinle Formation. Photo courtesy of the National Park Service.

23 Sequence of Triassic Bearing Formations.

	COLORADO	ARIZONA, NEW MEXICO			TEXAS
UPPER TRIASSIC		Kayenta			
		Moenave	Springdale		
			Dinosaur Canyon		
		Wingate	Lukachukai		
			Rock Point		
	Dolores	Chinle	Church Rock		Redonda
			Owl Rock		
				Correo	
			Petrified Forest		Chinle
			Moss Back	Poleo	Santa Rosa
			Monitor Butte	Salitral	Pierce Canyon
			Shinarump	Agua Zarca	

(Dockum spans the Texas column group)

3

The Jurassic Period

Dinosaurs of the Morrison Formation

Seas dominated the earth during the early and middle Jurassic period, leaving only isolated subcontinent-sized islands to nurture developing reptilian stocks. For most of this period great seas covered much of the present southwestern United States, and it was not until quite late in the Jurassic that significant continental sediments were deposited. Good Upper Jurassic Morrison exposures occur over large portions of northern New Mexico, but for reasons remaining unclear, vertebrate fossils are rare. Perhaps the local environment, being too hot, cold, or wet, could not support a large reptile population. Scant but sufficient evidence exists in New Mexico's Jurassic formations to hint of isolated dinosaur populations. Several identifiable bones were recovered from three Morrison localities west of Albuquerque and near Grants, including a claw and forelimb belonging to a giant carnosaur; the bones of a giant *Brontosaurus;* and the remains of *Stegosaurus.* These Jurassic dinosaur species, so rare in New Mexico, are common members of the Morrison fossil assemblage in Colorado and Utah. Jurassic rocks exposed at Garden Park, near Canyon City,

24 Carnosaur Footprint. Illustration by John C. McLoughlin.

25 Jurassic Rock Sequence Chart. In the south-western United States the Lower Jurassic Navajo Sandstone is a strikingly impressive formation of cement-hard white sand dunes in which only a few lightly built dinosaurs have been found. Vertebrate fossils in North America's Middle Jurassic are almost nonexistent, found only in isolated areas above a great inland sea. By late in the period, dinosaur populations exploded as the sea drained.

		NORTH AMERICA
Upper Jurassic		MORRISON
Middle Jurassic		
Lower Jurassic		
		NAVAJO

and at Morrison, Colorado, have produced gigantic dinosaur bones in abundance.*

In the Uinta Basin in northeastern Utah, near Jensen, Uinta County, an 80-acre tract of land has been set aside as part of the national park system under the name of Dinosaur National Monument. This protected area, as the name of the park implies, has within its boundaries a rather extensive series of dinosaur-bearing sediments.

The history of the Dinosaur National Monument begins in 1909 with the discovery by Earl Douglas of the Carnegie Museum in Philadelphia, of a series of well-preserved dinosaur skeletons. Between 1909 and 1915 such an abundance of fossil remains was found that in 1915, the secretary of the interior, at the urging of Dr. W. J. Hollard, director of the Carnegie Museum, withdrew this acreage from public domain and established it as a national monument to preserve its fossil resources. Excavation at Douglas' quarry continued up to the end of 1922 when the operation was abandoned.

For Earl Douglas the idea of closing down this great quarry was unthinkable, and his thoughts turned to exhibiting these large bones in place, chipping the matrix away on only one surface thus leaving the bones in relief against a frame of supporting sandstone. For various reasons, the government spent the next

---

*Of the more than 150 varieties of land animals and plants which have been discovered in rocks of Jurassic Morrison age, including two dozen kinds of primitive mammals, half of that list would be dinosaurs.

The following list of dinosaur genera which lived during Morrison times represents only the best known forms. There will, without any doubt, be many additions to this list as future discoveries are made.

Amphicoelais	Ceratosaurus	Hoplitosaurus
Antrodemus (Allosaurus)	Coelurus (Ornitholestes)	Hypsirophus
Apatodon	Creosaurus	Laosaurus
Apatosaurus (Brontosaurus)	Diplodocus	Maceolgnathus
Atlantosaurus	Diracodon	Morosaurus
Barosaurus	Dryosaurus	Pleurocoelus (Astrodon)
Brachiosaurus	Dryptosaurus	Stegosaurus
Brachyrophus	Dystrophaeus	Symphyrophus
Camarasaurus	Elosaurus	Tichisteus
Camptosaurus	Epanterias	
Caulodon	Haplocanthosaurus	

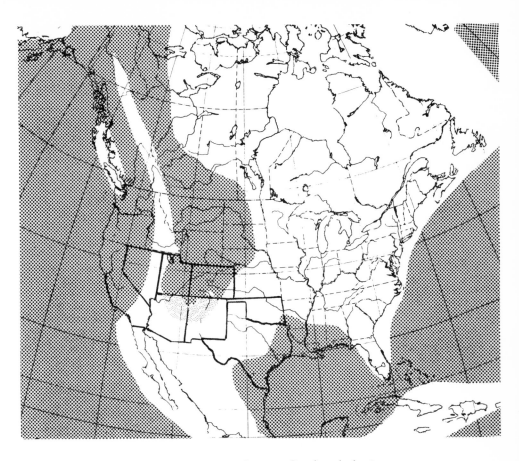

26 Early Jurassic Seas began covering part of western Canada and what is now the United States northwestern coast. Great sand dunes covered much of the Southwest. By the Middle Jurassic, this sea, called the Sundance because its sediments sparkled with gypsum crystals, had spread over parts of New Mexico, Arizona, Colorado, and Utah. By the Late Jurassic, it had been reduced to a giant swamp. As geological uplifts drained its water, thick deposits of mud, sand, and gravel remained to form the Morrison, Navajo, and Kayenta formations, entombing an amazing record of land animals.

thirty-five years "thinking" about Douglas' new idea. It wasn't until 1953 that another Dinosaur National Monument project got underway.

From 1935 to 1940, the National Park Service carried out extensive excavation at the site of Carnegie's original quarry, which today constitutes one of the most astounding and instruc-

27 National Park Service employee Jim Adams reliefs the skull of *Camarasaurus* on the face of the Dinosaur National Monument Quarry in northeastern Utah. Giant plant-eating dinosaurs like this *Camarasaurus* lived during the Jurassic period, about 140 million years ago. Photo courtesy of the National Park Service.

28 National Park Service paleontological technician Tobe Wilkins cleans up the skull of an *Antrodemus* in the laboratory of Dinosaur National Monument. *Antrodemus* was a carnivorous dinosaur that roamed the country about 140 million years ago. Photo courtesy of the National Park Service.

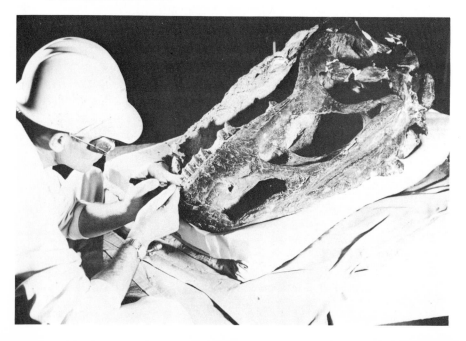

29 National Park Service paleontological technicians Adams and Wilkins have spent the last eighteen years exposing and reliefing the fossilized bones of fourteen different kinds of dinosaurs and five other prehistoric reptiles on the tilted face of the Dinosaur National Monument Quarry. To date over a thousand separate bones representing about twenty-five individual animals have been uncovered and catalogued. Photo courtesy of the National Park Service.

30   The tilted beds of the Morrison Formation (the fossil-bearing strata) can
be clearly seen just east of the unique dinosaur quarry and Visitor Center at
Dinosaur National Monument. Photo courtesy of the National Park Service.

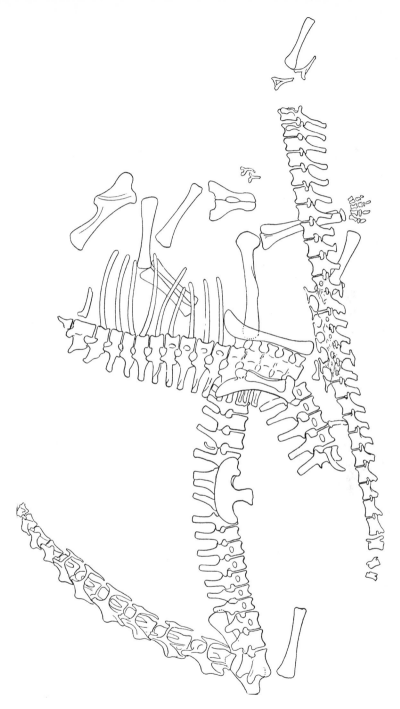

31 Sketch of *Diplodocus* skeleton, at Carnegie Quarry, now Dinosaur National Monument.

tive sites imaginable. Each year, hundreds of thousands of visitors come to the monument to marvel at the exposed dinosaur remains continually being excavated by highly trained Park Service personnel, skilled in both mining and stone cutting.

Despite the 300 tons of dinosaur bones removed during the early Carnegie Museum expeditions, many articulated skeletons of both large and small dinosaurs remain in the strata. In the quarry there is a veritable Noah's Ark of animals from the Upper Jurassic Morrison period. Found in a heavy, green, conglomeratic, cross-bedded sandstone are the bones of the largest of the giant sauropods, mixed with bones of smaller, but powerful carnivorous forms. Also there are bones of the armored *Stegosaurus*, as well as skeletons of the lightly built birdlike dinosaurs. Some skeletons found in the quarry are complete, with most of the bones articulated to one another. Other specimens are represented by only a portion of a skeleton or by an isolated series of bones. The diversity of animals represented within the boundaries of Dinosaur National Monument makes the site the most remarkable deposit of Upper Jurassic fossil reptiles yet discovered.

The nature of the sediments indicates the presence of an old sand bar deposited at the inside of a river bend, which, in its shallow waters, arrested the more or less decomposed carcasses as they drifted downstream toward it. Thus many different animals were brought together. The stranded carcasses there were rapidly covered by sand and held in place before decomposition of the ligamentary attachments allowed the bones to shift out of position. That many of the larger skeletons were not completely covered immediately is demonstrated by the fact that bones on the upper side often show greater displacement. That this scattering of the skeletal parts was due to current action is indicated by the bones themselves. Whenever shifting has taken place they will invariably be found to the east of the main portion of the skeleton. In other words, the direction of the current was from the west toward the east. Current action is further indicated by the character of the Morrison sediments in which the bones are embedded; that is, by the strong cross-bedding and the sorting of the fine and coarse materials of which the sandstone is composed.

Brightly colored strata of this Morrison formation were deposited over a forty-million-year span during the middle of the Age of Reptiles. In Utah and western Colorado, Morrison age strata accumulated to a thickness of over 900 feet, as layers of sandstone, shale, gypsum, and limestone were deposited within the swampy and seemingly endless lowland lakes and rivers which covered large portions of a poorly drained landscape. This was an ideal environment for the development of many unique varieties of plant and animal life. The dimly lit world beneath a tangled cover of dense ferns, so typical of this age, was alive with infinite numbers of tiny lizards scurrying from frond to frond. The reptiles' struggle for existence must have been incredibly fierce. For every tiny reptile there was a counterpart, just a little larger and a little faster, trying to catch and eat it. There was practically no end to this nightmarish pecking order, for the largest "hen" was a sixteen-thousand-pound monster.

Too Heavy for Land

Order—Saurischia
Suborder—Sauropoda
A) Family—Cetiosauridae
    Genus—*Diplodocus*
B) Family—Brachiosauridae
    Genus—*Apatosaurus* (also Brontosaurus)
    Genus—*Brachiosaurus*
    Genus—*Camarasaurus*
    Genus—*Alamosaurus* (Cretaceous of New Mexico)

The giants among the giant dinosaurs were the sauropods. Once thought to have been the largest of the land vertebrates, species of sauropods like *Brachiosaurus, Diplodocus,* and *Brontosaurus* (recently renamed *Apatosaurus*), are now considered to have been completely, or almost completely, water animals. Their tremendous weights—sometimes more than fifty tons—would

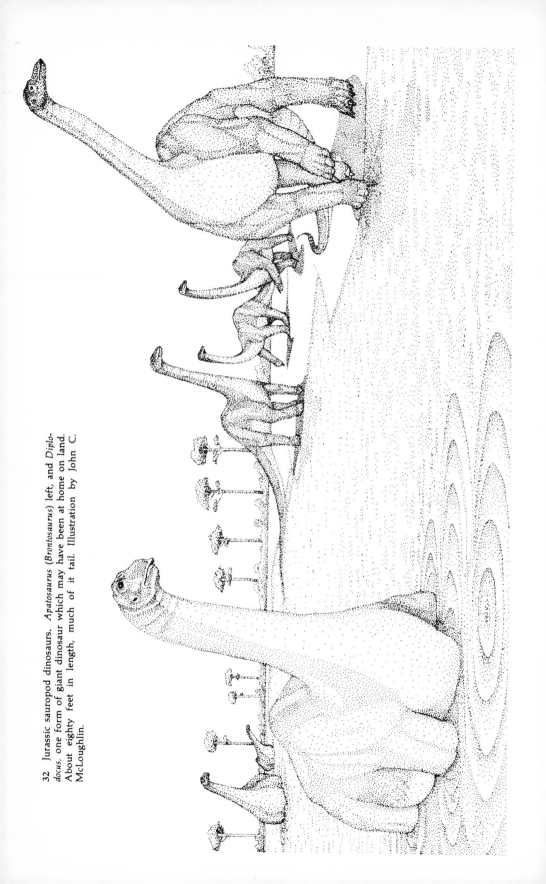

32 Jurassic sauropod dinosaurs. *Apatosaurus* (*Brontosaurus*) left, and *Diplodocus*, one form of giant dinosaur which may have been at home on land. About eighty feet in length, much of it tail. Illustration by John C. McLoughlin.

surely have damaged vital body structures had they ventured onto dry land for very long. The ponderous sauropods probably never mired, as did the carnosaurs, when taking their infrequent walks over marshy mud flats. Their feet had the ability to expand, or flatten, when weight was placed on them, and contract when the pressure was removed, as do the feet of modern elephants. In this way, the animal could literally sink up to its belly in mud and quite easily remove its feet from the holes. This ability surely saved many of these giants from an early end.

Sauropods were entirely dependent on soft, lush, swamp vegetation for their food. Their small jaws and weak front teeth were probably used to pull up tender plant shoots from both above and below the water. One may wonder how an animal of this enormous bulk with such a small mouth could gather enough food, mostly undigestable cellulose, to keep from starving. The explanation may be that such slow-moving giants, operating with the low metabolic rate of all reptiles, probably needed less food than a large modern mammal like the elephant, a considerably smaller, but much more active animal. If the reader frequents his local zoo and is familiar with its reptiles, he knows how much time these creatures spend simply lying around in the sunlight. Reptiles can move quite rapidly but seldom have the wish to do so. When they do find a need to exert themselves, they quickly become exhausted and must recharge their bodies' energy cells by resting for a long while. It must be assumed that dinosaurs acted in much the same way.

A large sauropod, like *Apatosaurus*, spent nearly all its life bobbing in deep water, while its head and neck served it in much the same manner as a trunk does an elephant while feeding. The sauropod's head was set on the end of a very long serpentine neck which made the animal tall enough to have looked over the top of a modern three-story building, or dredge plants from twenty or thirty feet below the surface of a lake. In many varieties, the nostrils were located on the crest of a small head, permitting the sauropod to remain completely submerged when frightened by a passing carnivore. This posture may have also been assumed for

33 *Brachiosaurus*, the giant of the sauropods, probably lived an aquatic life. This monster breathed through an air hole on top of its head, apparently an adaptation to life in deep water. Even though *Brachiosaurus'* tail was short, its body measured more than eighty feet in length and weighed close to fifty tons. Illustration by John C. McLouglin.

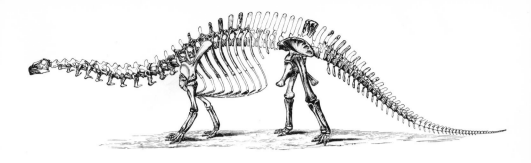

34 *Brontosaurus* skeleton. Restoration from Marsh.

purposes of resting, in the same manner as the hippopotami of
the Nile.

Commonly associated with sauropod remains are gastroliths, or
stomach stones. As much as a bushel of these variously sized
rocks, which represent stony material unknowingly ingested
along with plants, have been recovered from within the rib cages
of an *Apatosaurus.* These stones acted in place of molars, as a
digestive aid, grinding coarse plant fibers and hulls. Continually
rubbing against one another while in a suspension of stomach
acids and chewed-plant mulch has given the dinosaur's gastroliths
mirror smooth, slightly etched surfaces.

Sauropods of the Southwest's late Jurassic period varied greatly
in length, ranging from less than eighteen feet (*Camarasaurus*)
to eighty or a hundred feet. It is possible that some species similar
to *Diplodocus* were even longer. Unlike mammals, which grow
until reaching maturity, the reptiles, which include all dinosaurs,
increase their size at a diminishing rate throughout their life.
Barring any mishaps, like being killed or eaten, they could
theoretically have kept growing for several centuries.

A distinction is made between being killed and being eaten
because of the dinosaurs' regenerative capabilities. A large
portion of their anatomy, a complete limb or a tail, for example,
could be wrenched off the body by a predator and, if given
enough time, grow back, complete with new bones and a fresh
network of nerves. Modern reptiles which regenerate lost appendages usually end up only with a numb stump, a rather sorry
substitute for the original.

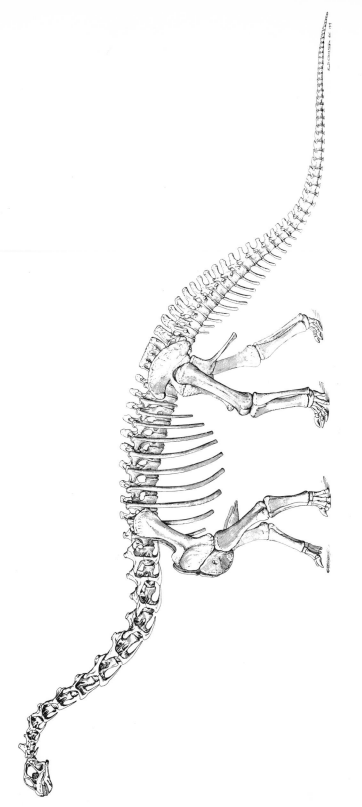

35 *Camarasaurus* skeleton. One of the smallest of the sauropods.

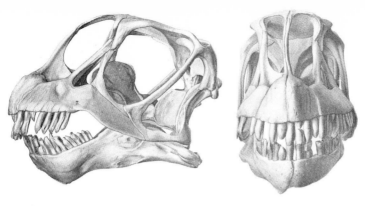

36 *Camarasaurus* skull. Notice the lightweight structure and the tiny vault for the brain. The teeth were used for grazing and were not adapted for grinding. From Osburn.

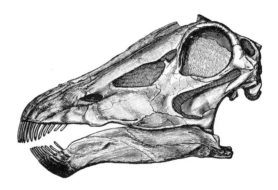

37 *Diplodocus* skull. One of the largest of the amphibious sauropods. Length of skull was about two feet; overall length of the dinosaur approached ninety feet. From Marsh.

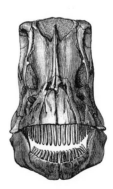

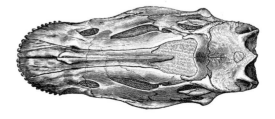

The Bird Thief

Order—Saurischia
Suborder—Theropoda
Family—Coeluridae
Genus—*Coelurus* (*Ornitholestes*)

The fossil remains of *Coelurus* (*Ornitholestes*) closely resemble those belonging to its Triassic forebear, *Coelophysis*. Its overall length only slightly surpassed that of the earlier coelurosaur, and much of it consisted of a long, stiff tail used as a counterbalance. Both of these two forms, as well as all other carnosaurs and many herbivores, were put together with their trunk region and head resting on their pelvis, well forward of center, which placed the center of balance at the acetabulum, or hip joint. If not for an equally weighty tail to balance the scale, the little dinosaur would have had difficulty standing upright. In classic therapod fashion, *Coelurus'* (*Ornitholestes*) angular skull housed a battery of small, sharp, double-edged teeth. Well-developed arms, claws, grasping fingers, and strong swift legs may have caused early students of dinosaurs to dub this creature "The Bird Thief," a slightly less than accurate description of its eating habits.

With the exception of small, unobtrusive lizards trying to keep from underfoot, these small archosaurian reptiles were without doubt the most agile land animals of their time. They had to be. Not only did their diet consist of birds, but more than likely *Coelurus* (*Ornitholestes*) mainly fed on the young of other dinosaurs. Obviously some of the other dinosaurs would have been upset by such kidnappings, and undoubtedly gave chase.

Coelurus (*Ornitholestes*), except for its dimensions, was a good example of a carnosaur type. Judging by the large size of eye sockets, nostrils, and ear cavities, the saurian predator must have had acute sense organs. Though its brain was no more than a pea-sized bit of grey tissue hidden near the back of its skull, instincts played a major role in the life of this, and all dinosaurs. Instincts are inherited, unlearned, programmed responses to environmental stimuli. The little dinosaurs were born knowing

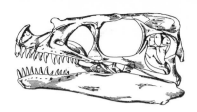

38 *Ornitholestes* skull. About eight inches long. After Osburn.

precisely how and what to eat, who their enemies were, how to reproduce, but not, unfortunately, how to adapt to changing environments. Since early dinosaur ancestors did not encounter monumental stresses, this critical bit of information was never stored in the animal's genetic makeup. The dinosaurs thus had either to change biologically, or perish, and the former was out of the question.

Plates and Spikes

Order—Ornithischia
Suborder—Stegosauria
Family—Stegosauridae
Genus—*Stegosaurus*

One of the more bizarre forms of Jurassic dinosaurs was a moderately sized creature called a stegosaur. *Stegosaurus*, the most familiar species in this biological suborder, reached a maximum length of about twenty feet. What made it unusual was the fact that its vertebral column was protected by a double row of twenty-six large, erect, horn-covered plates which were embedded, as with the armor of most dinosaurs, in the upper layers of its tough leathery skin. These bony plates may have acted as much as a psychological as a physical defense mechanism, making *Stegosaurus* seem at least a little less appetizing to larger contemporary carnivorous dinosaurs. If not, the *Stegosaurus* had a tail which he could thrash about with great force, and which was equipped with four long, heavy spikes capable of inflicting a great deal of damage.

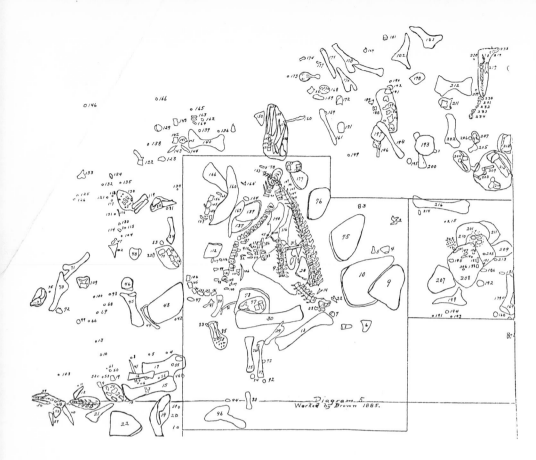

39 Page from a dinosaur collector's notebook. Barnum Brown of the
American Museum sketched this deposit of *Stegosaurus* bones in 1886.

Stegosaurus probably subsisted on soft vegetable material that it
could browse for in wooded areas. Not only were forests valuable
sources of food for these dim-witted monsters, but they also
afforded additional protection against attack. *Stegosaurus'* sides
were unprotected except for small bony platelets, a panoply of
ossicles, imbedded within its thick hide. Its belly, once thought
to be covered only with thick skin, is now known to have been
protected by slender, rodlike splints directly under, and originat-
ing from, the skin. In an x-ray, this arrangement would have
looked much like a Sioux breastplate of latticed bone tubes.
Stegosaurus, built as low to the ground as it was, would have
found these gastralia necessary to keep from gashing open its

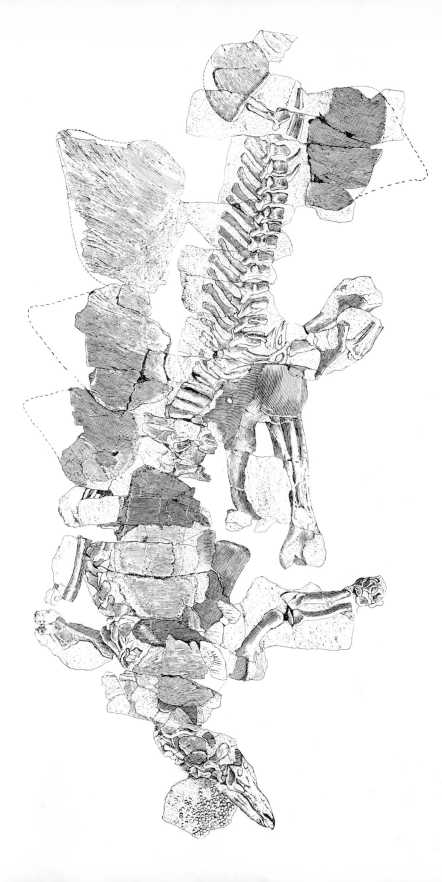

40 *Stegosaurus* skeleton, as found. Often years are needed to collect, prepare, and finally restore a dinosaur skeleton. Notice the pile of small round bones near the dinosaur's neck at the far left. These represent nodules of armor which protected its throat. After Gilmore.

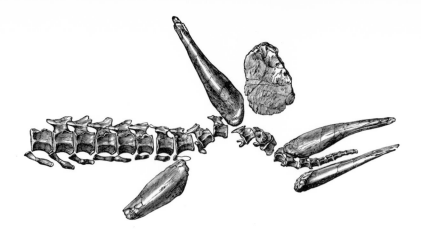

41 Tail of *Stegosaurus* showing the position of defensive spikes, as found. From Marsh.

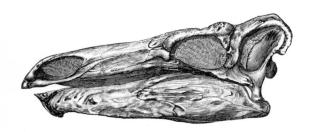

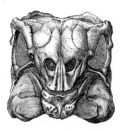

42 Skull of the Jurassic plated dinosaur *Stegosaurus*. Length about sixteen inches. From Marsh.

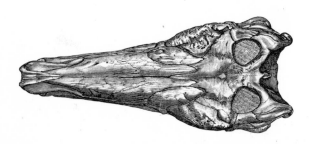

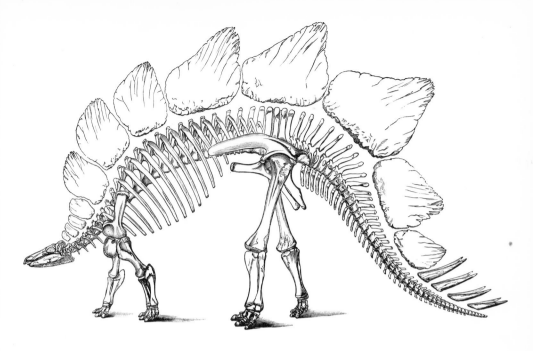

43 Skeleton of *Stegosaurus,* a Jurassic plated dinosaur. From Marsh.

44 *Stegosaurus,* a large armored dinosaur from the Upper Jurassic. Because of the animal's tiny skull and weak jaws, paleontologists believe that its diet consisted of soft upland vegetation. *Stegosaurus* must often have fallen prey to carnosaurs such as *Antrodemus,* the skulls of which can be seen here. Photo courtesy Royal Ontario Museum.

belly on sharp rocks. Carnivorous forms, like *Gorgosaurus* (*Alberto-saurus*) whose belly was exposed at all times, needed this protection because of its perpetually violent life. Another need for gastralia was as support for the gastric tissues. A dinosaur's full stomach was quite heavy, and the pressure it exerted upon the huge surface of belly skin would have been disastrous without some form of support. Constant hernias and ruptures of its digestive tract, or, at the very least, a stomach dragging along the ground in quadrupedal forms, would have been the rule.

The skull of *Stegosaurus* was tiny, and the brain, which was about the size of a kitten's, weighed no more than two and a half ounces. Reptilian brain matter, unlike that of mammals, is incapable of complex thought processes and rapid interpretation of environmental stimuli. Out of necessity, therefore, many dinosaurs like the stegosaurs developed additional relay stations from which nerve impulses could operate various parts of their bodies, rather than depending upon a feeble brain twenty to fifty feet away. At the base of *Stegosaurus'* tail was one such bulbous enlargement of the spinal cord. When this "second brain" was first discovered, newspaper reporters quickly picked up the story, twisted the facts in a minor way, and christened *Stegosaurus* as the dinosaur with two brains. In 1912, Bert Leston Taylor, a columnist on the staff of the *Chicago Tribune* portrayed *Stegosaurus* in his fanciful poem, which is quoted below.

The Dinosaur

Behold the mighty dinosaur,
    Famous in prehistoric lore,
Not only for his power and strength
    But for his intellectual length.
You will observe by these remains
    The creature had two sets of brains—
One in his head (the usual place),
    The other at his spinal base.
Thus he could reason "Apriori"
    As well as "Aposteriori."

No problem bothered him a bit
    He made both head and tail of it.
So wise was he, so wise and solemn,
    Each thought filled just a spinal column.
If one brain found the pressure strong
    It passed a few ideas along.
If something slipped his forward mind
    'Twas rescued by the one behind.
And if in error he was caught
    He had a saving afterthought
As he thought twice before he spoke
    He had no judgment to revoke.
Thus he could think without congestion
    Upon both sides of every question.
Oh, gaze upon this model beast,
    Defunct ten million years at least.

Stegosaurs were the first major dinosaur group to disappear, becoming extinct not long after the close of the Jurassic period, some 125 million years ago.

Teeth and More Teeth

> Order—Saurischia
> Suborder—Theropoda
> Family—Allosauridae
> Genus—*Antrodemus* (also *Allosaurus*)

Without question, the most savage creatures to have roamed North America were the giant carnosaurs: the meat-eating dinosaurs from the Age of Reptiles. *Antrodemus* (*Allosaurus*), a typical carnosaur species of the Jurassic period, grew to lengths of thirty-four feet, and weighed more than four tons. Its head and

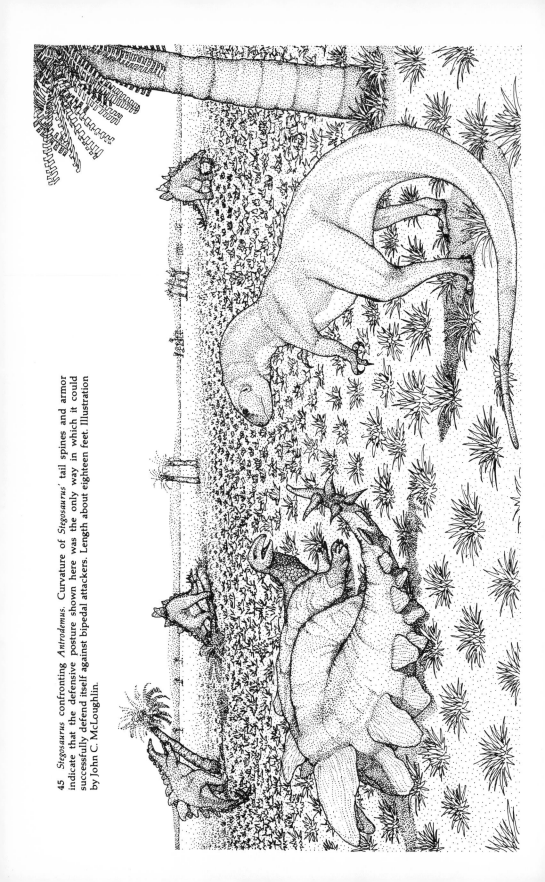

45 *Stegosaurus* confronting *Antrodemus*. Curvature of *Stegosaurus'* tail spines and armor indicate that the defensive posture shown here was the only way in which it could successfully defend itself against bipedal attackers. Length about eighteen feet. Illustration by John C. McLoughlin.

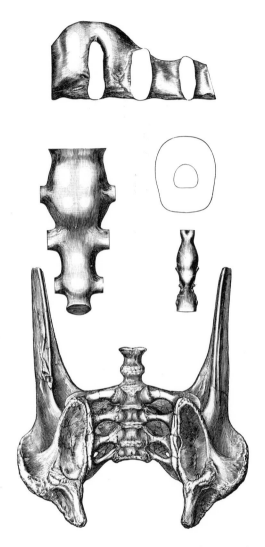

46 Full-scale *Stegosaurus* brain, about the size of a walnut. It weighed two and a half ounces. The animal weighed ten tons. The reptilian nervous system transmits messages only a few yards per second. If a large sauropod, like *Diplodocus,* had the tip of its tail bitten, the brain, some eighty-seven feet away, would not have registered pain for the better part of a minute. An enlargement of the spinal cord in the hip region took over many brain functions in the operation of the animal's hind quarters.

47   Skeleton of *Antrodemus* (*Allosaurus*).   Giant carnivorous dinosaur from the Jurassic of Utah. Photo courtesy Royal Ontario Museum.

48   Skull of *Antrodemus*.   The loosely jointed construction of the carnosaur's skull made it possible to bolt down large pieces of meat in the same way a modern snake can separate the bones in its skull to swallow large prey. Photo courtesy Royal Ontario Museum.

49   *Antrodemus,* better known as *Allosaurus* ("jumping lizard"), was an agile and active dinosaur during the Upper Jurassic. Photo courtesy Royal Ontario Museum.

50   Tooth of the Jurassic carnosaur *Antrodemus,* found west of Albuquerque. Photo courtesy Tucson Mineral and Gem World.

jaws were disproportionately large when compared to its body, a logical adaptation in light of its predatory role in the dinosaur world. Structures on the skull of *Antrodemus* indicate that it may have been able to expand even further its already monstrous mouth by disjointing certain bones in the jaw and the skull's frontal area. This curious ability enabled it to bolt down massive unchewed chunks of flesh—possibly even an entire animal. Some modern snakes have this same ability and are quite capable of swallowing whole animals many times the diameter of their own bodies.

The *Antrodemus'* hind limbs were extremely strong and gave these large reptiles their great speed. Each foot had four huge toes, all equipped with sharp, curved claws. These were used to hold victims while the great mouth with its arsenal of long recurving teeth tore a carcass to pieces. The front feet, or more precisely arms, were smaller and though they retained little of their ancestral grasping function, were not altogether useless.

Bones of the giant aquatic sauropods have been uncovered which show *Antrodemus* teeth marks, and the knifelike teeth themselves, apparently broken off in a death struggle, are often found as well.

Order—Saurischia
Suborder—Theropoda
Family—Ceratosauridae
Genus—*Ceratosaurus*

Ceratosaurus was another large carnosaur, a contemporary of the *Antrodemus* and other Jurassic deinodont dinosaurs. Its build was similar to that of *Antrodemus,* but only half its length. *Ceratosaurus,* from the Greek words meaning horned lizard, was unique among carnosaurs because it possessed a nasal horn and bony shields

51 A pack of allosaurs surrounds a *Diplodocus*, cutting it off from the water. Although one allosaur probably could not have brought down such large prey, a group did so without trouble.

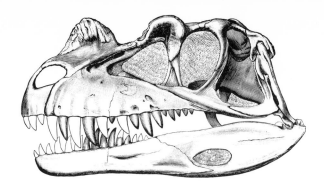

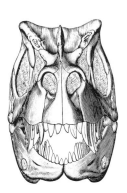

52 Skull of the Jurassic carnosaur *Cerato-saurus*. Length about two and a half feet.

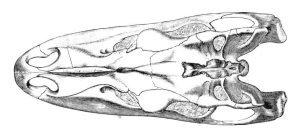

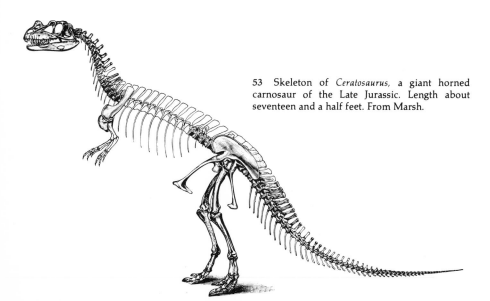

53 Skeleton of *Ceratosaurus*, a giant horned carnosaur of the Late Jurassic. Length about seventeen and a half feet. From Marsh.

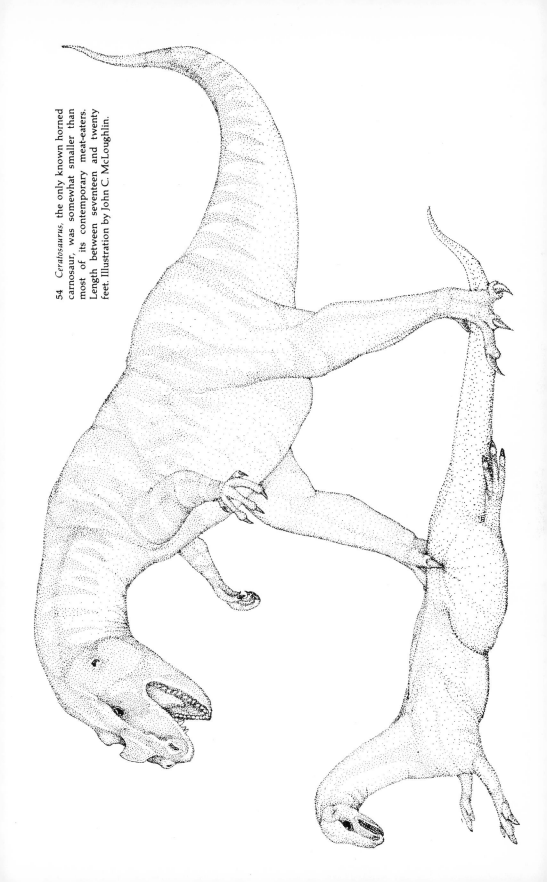

54 *Ceratosaurus*, the only known horned carnosaur, was somewhat smaller than most of its contemporary meat-eaters. Length between seventeen and twenty feet. Illustration by John C. McLoughlin.

above each eye. Unlike the conical horns of the later Ceratopsians, this horn was more bladelike, and considerably smaller. The horns may indicate *Ceratosaurus'* possible habit of shoving heavy brush aside.

The feet of *Ceratosaurus* differed from those of other carnosaurs in that all three metatarsals had fused into one solid bone, much like the foot of a bird. This seems a strange adaptation for a large predator, since, with the instep bones fused, much of the flexibility and bounce would be gone from the animal's stride. It would slow the animal down and produce a more birdlike strut when in pursuit of prey.

4

The Cretaceous Period

Of the three Mesozoic periods the Cretaceous was the last, and had in many respects the most dramatic and varied life-forms. Its name derived from the Greek word meaning chalk because the first description of this age came after discoveries in the highly fossiliferous marine chalk cliffs of Dover, England. A good deal of attention has been given to Cretaceous environments, for within their long history of orogenies and catastrophies the reptilian populations flourished beyond comprehension. Grassless tropical jungles covered much of today's southwestern desert, in an abundance clearly illustrated by endless deposits of decomposed, carbonized vegetable matter, or coal, many of which are hundreds of feet thick. Finely bedded shale in north central New Mexico and Colorado have preserved a detailed record of this flora in the form of carbonized leaves and a few delicate flowers. Using these fossils as guides, we can clearly visualize a Late Cretaceous landscape.

Within the dense swamps were the common Ficus, water lilies, horsetail rushes, eucalyptus, and many species of trees similar to cypress. On higher, better drained ground were grapes, laurels, walnut trees, ivies, oaks, sequoias, ash, bayberry, beans, ebony, ferns, fir, honeysuckle, palms, poplar trees, willows and many others. It was no small wonder that plant-eating dinosaurs were abundant, as were the huge flesh-eating forms which preyed upon them.

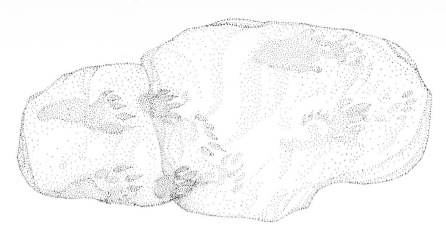

55   Mammal Tracks.   Illustration by John C. McLoughlin.

As one studies the preserved Cretaceous dinosaurs and other reptiles of their time, it is often possible to trace the most minute evolutionary transformations beginning at the bottom, or oldest, of the rock sequences, and proceeding to the higher, and relatively younger ones.

In the southwestern states, this sequence of rocks has been correlated to those already dated in the better known northern quadrant of North America. The following geologic chart exhibits this correlation and similar relationships with other Cretaceous dinosaur-bearing formations.

The overwhelming majority of Upper Cretaceous dinosaurs were herbivorous. In fact, they outnumbered the predaceous, carnivorous dinosaurs by about four to one, a predictable figure considering the predator-to-prey ratio of modern animals. The numbers of both types of Cretaceous dinosaurs during this period staggers the imagination. "Herds" of plant-eating giants swarmed over the marshy landscapes, much like the American bison whose sheer numbers blackened western prairies.

Giant carnivorous dinosaurs became an ecologic necessity. As the African lion unknowingly acts to control populations and retard the spread of disease by sick animals, so did the carnosaurs. The very young and the old, sick, or crippled dinosaurs were easy prey. With such an abundant food source, there was seemingly no limit to the size of the methuseloid hunters.

56   Cretaceous rock sequence.

57 *Gorgosaurus* (*Albertosaurus*) and its prey, a duck-billed ornithopod dinosaur of the Upper Cretaceous. Bones have been found all the way from southern Arizona to Canada. About twenty-nine feet in length. Illustration by John C. McLoughlin.

58 Skull of *Gorgosaurus*. Openings indicate that this reptile had acute senses of sight and smell. Photo courtesy Royal Ontario Museum.

59 *Gorgosaurus* (*Albertosaurus*), a huge deinodont carnosaur from the Upper Cretaceous. Notice the gastralia (abdominal ribs) which many museums fail to include on mounted skeletons. Photo courtesy Royal Ontario Museum.

60 Footbones of *Gorgosaurus* (*Alberto-saurus*) adapted, as are those of birds, for grasping prey. Photo courtesy Royal Ontario Museum.

Beside the vast numbers of dinosaurs during the Mesozoic, there existed great populations of reptiles, many of which were so tiny that they could exist unnoticed underfoot in today's southwestern desert. Other giants, which were not dinosaurs, also inhabited the area. These included a giant crocodile, *Phobosuchus*, from the Big Bend region of Texas, which grew to fifty feet in length and most likely fed on immature dinosaurs; and champosaurs which, with their long, pointed snouts and numerous needle-sharp teeth, ecologically filled the same niche as the Komodo "dragons" of present day Madagascar.

The mammal population grew in number and importance during the Late Cretaceous. Some forms, such as the opossum *Didelphis*, exist today practically unchanged, a true "living fossil." *Hesperornis* and *Ichthyornis*, birds with teeth but without the ability to fly, rode the waves of shallow seas. Giant marine lizards, the plesiosaurs and the mosasaurs, were relentless predators below.

Although environments of the Cretaceous were much like those of the Upper Jurassic, topographical changes, which would severely affect living conditions did occur. Continents began —slowly—to form, permitting large swampy basins to drain their waters into existing oceans. All of this may have caused the disappearance of many water-living dinosaur forms which were unable to adapt. Amphibious sauropod giants, the primary food

61 Skull of the giant crocodile *Phobosuchus,* with large modern crocodile in foreground. Photo courtesy The American Museum of Natural History.

source of the carnivorous species, were probably the critical link in a long dinosaur food chain. As they became extinct, so did the carnivores. Those animals which depended upon the eggs of the meat-eaters and the sauropods could not survive; those which ate the egg-eaters starved; and so on. With so many species dependent on each other, it is not surprising that the total extinction of the dinosaurs came so rapidly—in a geologic instant—at the close of the Cretaceous.

The Tyrant King and His Court

Largest and most impressive of the giant carnosaurs, or meat-eating reptiles, was the "Tyrant King," *Tyrannosaurus rex.* This name is most apt, for no other animal, living or extinct, has ever matched the awesome power and ferocity of this species.

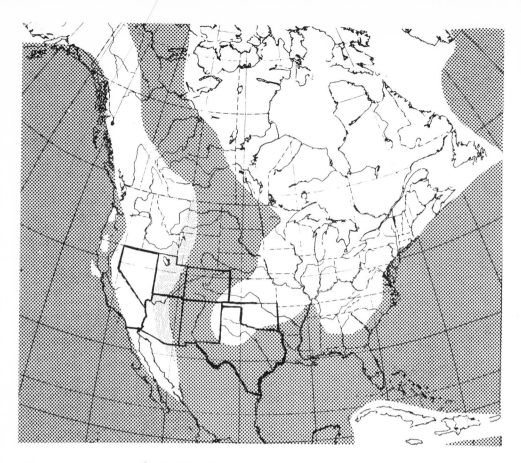

62 The North American continent during the Cretaceous period. As great mountain systems rose in the far western states, much of today's Southwest was inundated by rising seas. Late in this period, with the birth of the Rocky Mountains, the seas began to drain, leaving a swampy lowland where great populations thrived.

Like the smaller *Antrodemus* of the previous geologic period, *Tyrannosaurus* and a similar Cretaceous giant, *Deinodon*, were extremely effective hunters. Massive beasts, their powerfully built hind legs may have carried a fifty-foot, sixteen-to-twenty-thousand-pound body fifty miles per hour, or faster.

The skulls of these Cretaceous monsters were large, even for animals of their size. The *Tyrannosaurus* skull measured more than four feet in length. Many of the sixty teeth in its gaping mouth were over six inches long, and all curved backwards toward the throat as an aid in swallowing great chunks of meat. (It has been calculated that these large carnosaurs could swallow

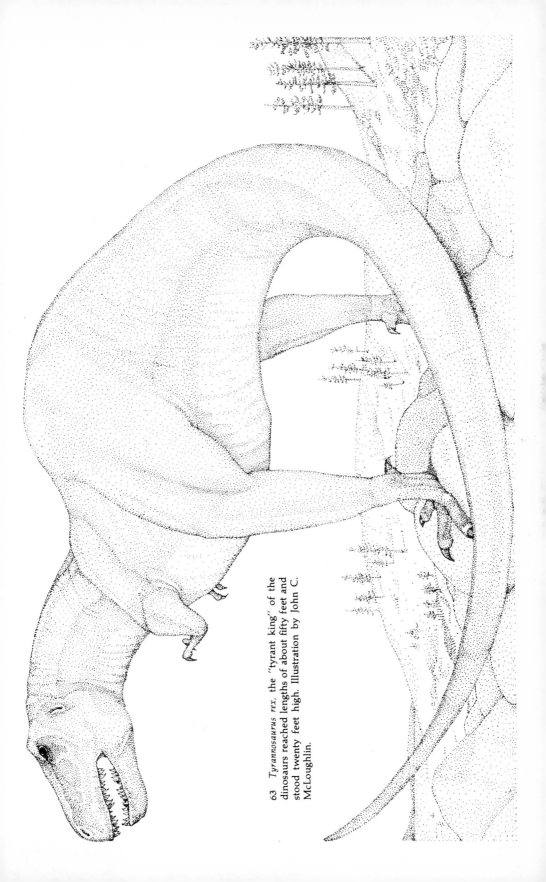

63 *Tyrannosaurus rex*, the "tyrant king" of the dinosaurs reached lengths of about fifty feet and stood twenty feet high. Illustration by John C. McLoughlin.

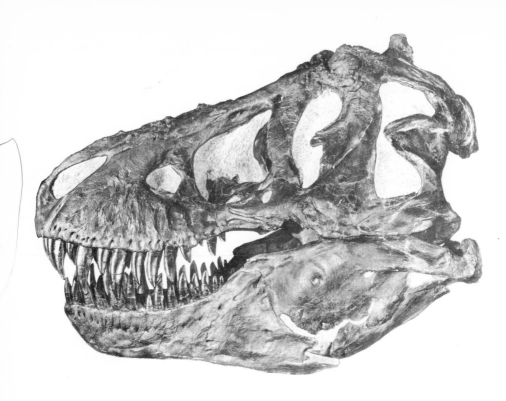

64 Skull of *Tyrannosaurus rex*. About fifty inches in length. From Osburn.

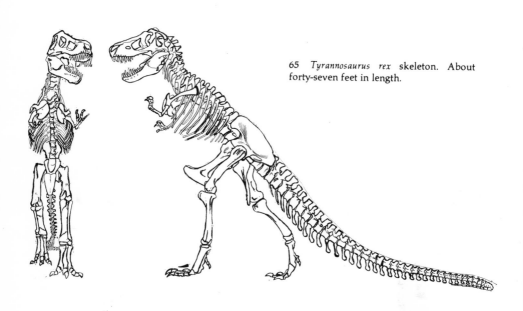

65 *Tyrannosaurus rex* skeleton. About forty-seven feet in length.

more meat in one bite than a contemporary family of three humans could eat in several months.) Furthermore, its teeth were serrated, which enabled these predators to tear through the leathery skin and dermal armor of their unfortunate prey.

The front limbs of Late Cretaceous deinodonts (meaning terrible teeth) were reduced in size, being mere stubs with little or no remaining function. They were much too short to reach the mouth and surely too weak to aid in capturing, rending, or even holding prey. *Gorgosaurus* (*Albertosaurus*), a relatively small Late Cretaceous carnosaur of southern Arizona, except for its shorter length and proportionately lighter build, was a close copy of other deinodonts.

The fossil bones of carnivorous dinosaurs are reported to be more common in Cretaceous strata in New Mexico, Colorado, Texas, and Utah than in beds of the same age in the northern regions of the United States. One particularly interesting discovery—indicative of the life cycle of the period—was made at Glen Rose, Texas. In lower Cretaceous limestone, representing an ancient beach, the tracks of a great sauropod could be seen making their way across the mud. Superimposed on these are the footprints of a large carnosaur clearly on the hunt. When the sauropod tracks turn to one direction, so do those of the carnosaur; its own clawed feet stepping directly onto the deeper prints of its huge prey.

"Ostrich" Dinosaurs

Order—Saurischia
Suborder—Theropoda
Family—Ornithomimidea
Genus—a) *Struthiomimus*
       b) *Ornithomimus*

Struthiomimus, and the related saurischian form *Ornithomimus*, were dinosaurs having the general proportions of a large bird—very long, slender hind legs, a long neck, and a small head. An incomplete skeleton might easily be mistaken for that of an

66 Footprint of *Tyrannosaurus* found in the roof of a coal mine near Horse Cañon, Utah. Next to it, for size comparison, is the author's two-year-old niece, Elizabeth Ann Grisham. The footprint specimen is in the Geology Museum, University of New Mexico, Albuquerque.

67 Over 100 million years ago in what was a soft beach of lime mud deposited near the present city of Glen Rose, Texas, a large sauropod was stalked by a smaller carnivorous dinosaur. The tracks of the carnosaur are seen in this photo paralleling the sauropod's trail and in a place or two fall directly on the sauropod's track. Photo courtesy The Texas Memorial Museum, The University of Texas at Austin.

68   *Struthiomimus,* a lightly built coelurosaur whose horny beak was adapted to eating eggs. Since dinosaurs may have been parentally protective, these ornithomimids were powerful runners, capable of making a swift retreat. Illustration by John C. McLoughlin.

ostrich. The hands of both forms were well adapted for grasping; the three inner fingers were long and the first seems to have been prehensile. This grasping capability, the creature's curious lack of teeth (functionally replaced by a horny beak), and great swiftness point to an egg diet. Well-developed hands would be needed to handle the eggs, teeth are not necessary to crush the shells, and swiftness would be needed to escape the anger of the eggs' rightful owners.

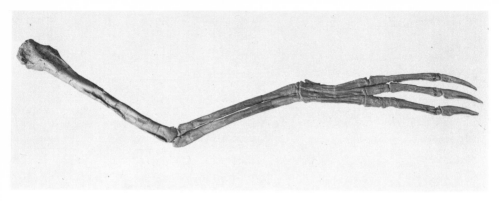

69　Grasping limb of *Struthiomimus*. Photo courtesy Royal Ontario Museum.

70　*Struthiomimus*, small Cretaceous birdlike dinosaur. Photo courtesy Royal Ontario Museum.

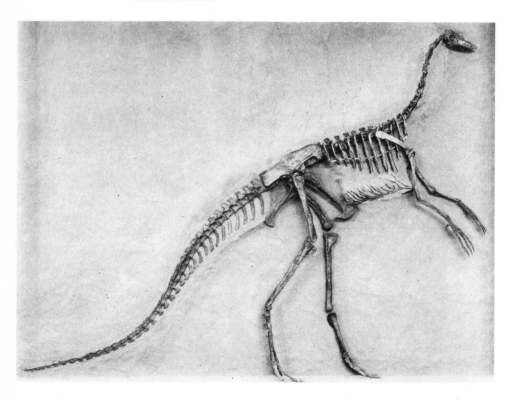

The Armored Dinosaurs

Order—Ornithischia
Suborder—Ankylosauria
Family—Nodosauridae
Genus—*Ankylosaurus*
    Nodosaurus
    Edmontonia
    Palaeoscincus
    Dyoplosaurus
    Panoplosaurus
    Scolosaurus
    Stegopelta
    Euoplocephalus

Ankylosaurs were great, lumbering, quadrupedal, ornithischian dinosaurs. They grew to be as long as their cousin *Stegosaurus*, but were considerably more massive. *Ankylosaurus*, the largest of the species, was protected by a series of thick, bony armored plates which completely covered the back surface of its body and tail. Its head was shielded by a mosaic of smaller, polygonal bony plates, supported by a dense construction of thick bone. The broad, flattened body and the head (which was spiked in some forms) resembled that of a horned toad. When attacked, the *Ankylosaurus* could simply lie down to protect its soft underbelly, and thrash its tail at whatever beast was molesting it. Unlike the armored turtle, however, *Ankylosaurus* was unable to retract its limbs. The end of its tail was equipped with heavy bone clubs or spikes which could easily cripple or kill a large carnosaur. In several species, long, sharp spikes at the edge of the armored plates may have discouraged the more aggressive animals from trying to flip them over. Oddly enough, the glyptodonts of the Pleistocene Ice Age fit this same description, although they were mammals.

Like all of the ornithischian dinosaurs, ankylosaurs were peaceful herbivores, which in the end proved to be their downfall.

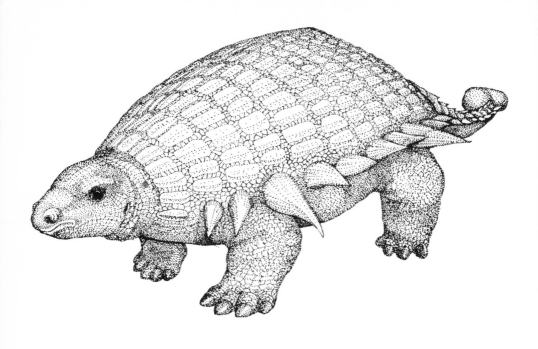

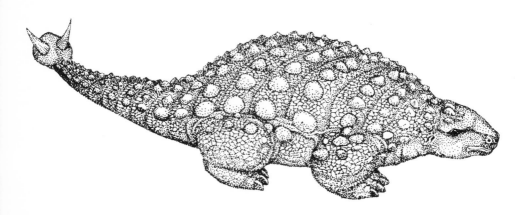

71 Two Upper Cretaceous armored ankylosaurs. Left is *Ankylosaurus;* right is *Palaeoscincus.* Both forms reached lengths of about seventeen feet. Illustration by John C. McLoughlin.

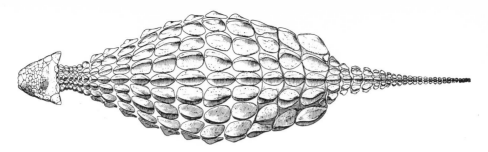

72 Protective armor of the Upper Cretaceous ornithischian dinosaur *Ankylosaurus*, seen from above. After Brown.

As the plants they ate disappeared with the rising of the continent at the end of the Cretaceous, so did these fascinating armored forms.

The Duck-Billed Dinosaurs

Order—Ornithischia
Suborder—Ornishopoda
Family—Hadrosauridae
Genus—*Corythosaurus*
  Parasaurolophus
  Kritosaurus
  Trachodon
  Claosaurus

Toward the end of the Cretaceous period, another curious giant reptile, the hadrosaur, or duck-billed dinosaur, developed. They derived their popular name from the broad flat construction of both the upper and lower jaws, which gave the appearance of the bill of a duck. Each side of the jaws held over 500 teeth, for a total of more than 2,000 in the entire mouth. Each of the small irregular teeth grew tightly against one another, overlapping in shingle fashion to form a nearly solid mass on which the animal efficiently ground the coarse vegetation it ate.

The largest species of hadrosaur, *Trachodon* (*Anatosaurus*) grew to thirty or forty feet in length, and weighed several tons. Its long hind legs were massive and, like most species of hadrosaurs, it

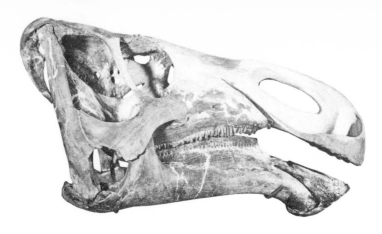

73 Skull of the "Roman-nose" hadrosaur *Kritosaurus navajovius*, found on New Mexico's Navajo reservation near Farmington.

walked upright. This extended the range of vision so that it could spot meat-eaters before being spotted. The toes were large, and from well-preserved fossil footprints on hardened beach sands, it was determined that there was a thick, leathery webbing between them.

Hadrosaur fossils are most often found near remains of aquatic animals like garfish, turtles, and crocodiles. This would seem to indicate that the duck-bills spent a great portion of their lives swimming in deep lakes and rivers, or at least wading in the shallows ready to flee to the safety of deeper waters if a predator approached. Strong muscles and tendons attached along the vertebrate column were used to move the tail from side to side and acted as a "propeller" when the animals did swim.

Worn teeth of hadrosaurs, having been pushed out and replaced by new dentition, are extremely common in Cretaceous river deposits. The author was able to pick up thousands of these diamond-shaped rejects from one small area, along with many fragments of calcified tendons.

Hadrosaurs living during the Southwest's late Cretaceous were named for the distinctive shapes of their skulls. *Corythosaurus'* skull was capped with a helmet-shaped crest; *Parasaurolophus'* crest was long and tubular; that of *Lambeosaurus* was shaped like a hatchet. The nasal bones of New Mexico's *Kritosaurus* were expanded, giving it a grand "Roman nose." All are classified as "duck-billed."

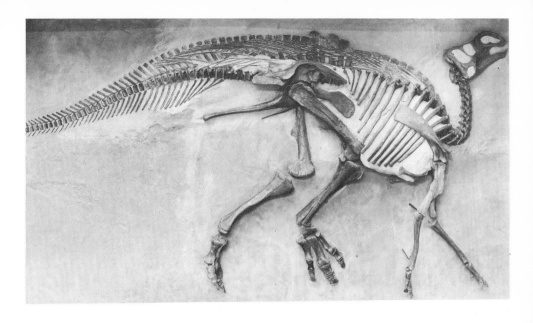

74 *Kritosaurus*, a hadrosaurian dinosaur from the Upper Cretaceous. Photo courtesy Royal Ontario Museum.

75 Early twentieth-century restoration of a possible Upper Cretaceous scene in northern New Mexico. Three duck-billed dinosaurs are shown: *Corythosaurus*, swimming; *Kritosaurus*, standing right; and *Trachodon* (*Anatosaurus*), standing left. Drawing by Richard Beckert.

76 Skull of *Parasaurolophus*. Photo courtesy Royal Ontario Museum.

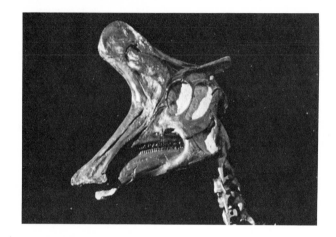

77 Skull of the crested hadrosaur *Lambeosaurus* from the Upper Cretaceous. Photo courtesy Royal Ontario Museum.

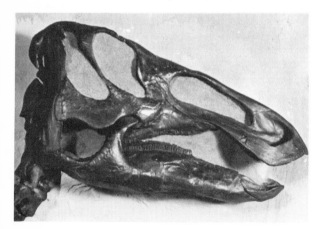

78 Skull of *Anatosaurus*, an Upper Cretaceous duck-billed dinosaur. Photo courtesy Royal Ontario Museum.

79 *Lambeosaurus*, a crested ornithopod hadrosaur from the Upper
Cretaceous. Footprints of these aquatic dinosaurs indicate that there was
thick leathery webbing between its toes. Photo courtesy Royal Ontario
Museum.

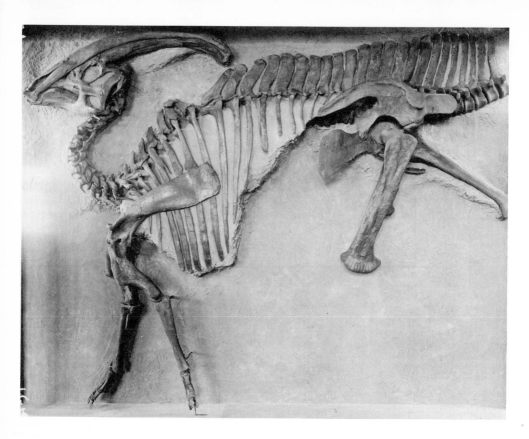

80 Skeleton of *Parasaurolophus*, a rare hadrosaurian dinosaur from the Upper Cretaceous. Photo courtesy Royal Ontario Museum.

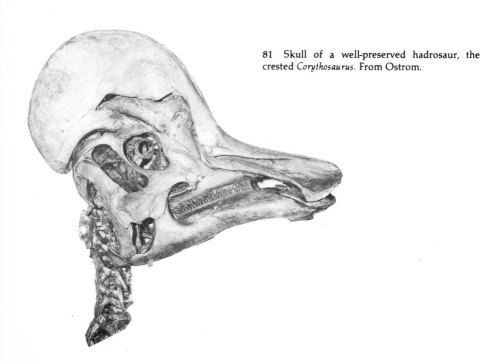

81 Skull of a well-preserved hadrosaur, the crested *Corythosaurus*. From Ostrom.

82 *Corythosaurus* skeleton. Photo courtesy Royal Ontario Museum.

Several hadrosaur "mummies" have been uncovered, one of which is on exhibit in the American Museum of Natural History, New York City. Highly detailed impressions of their rough pebbly skin are visible on the stone which surrounded the fossil (although no actual skin remains), and some tendons were fossilized. These rare specimens probably baked in the sun to such a point that their skin and muscles took on a leatherlike texture. Then they were buried by sand or clay before scavengers happened upon them. These unusual fossils also show the webbed toes and naked unarmored skin.

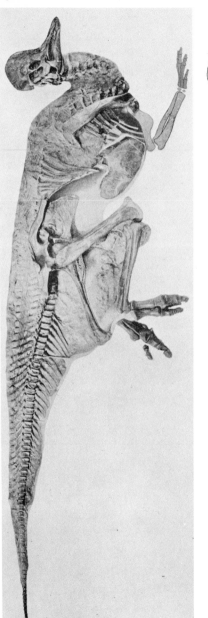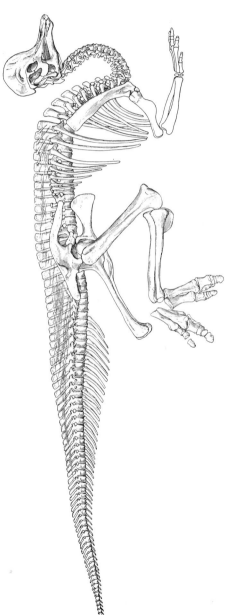

83 Mummy of *Corythosaurus*. Artist's detailed drawing (below) amplifies the bones and fossil tendons along the back and tail. From Brown.

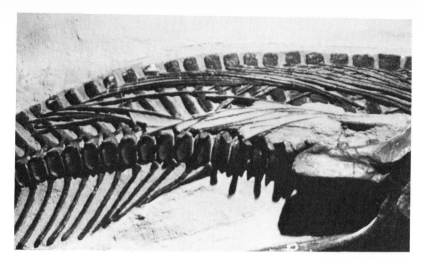

84 Tendons found preserved on the tail of *Kritosaurus*. Photo courtesy Royal Ontario Museum.

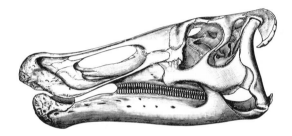

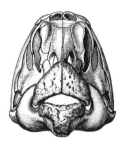

85 Skull of the duck-billed dinosaur *Claosaurus*. Hadrosaurs like this species used their bill to shovel up large quantities of the plant material they ate. From Marsh.

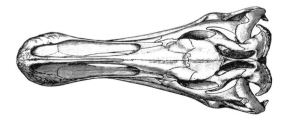

86   Restoration of *Claosaurus'* skeleton. About twenty-five feet in length. From Marsh.

87   Skeleton of *Prosaurolophus*, a hadrosaurian dinosaur from the Upper Cretaceous. Photo courtesy Royal Ontario Museum.

Bone-Heads

Order—Ornithischia
Suborder—Ornithopoda
Family—Troödontidae
Genus—*Pachycephalosaurus*
Troödon

The bone-headed relatives of the "duck-billed" hadrosaurs were an interesting lot. Inoffensive grazing reptiles, they subsisted on green land plants which they could nip off with their sharp front teeth. Similar in size to the duck-billed forms, instead of developing a thin hollow crest and flat bill they grew incredibly thick, solid domes of bone which extended as much as nine inches above their tiny brain cases. The need for such a thick skull is unclear, but it may have been used as a battering ram for defense, or in mock battles to impress a potential mate during the mating season. *Pachycephalosaurus,* a species whose face and head

88 *Pachycephalosaurus,* an Upper Cretaceous boneheaded ornithopod. The skull of this form is about twenty-six inches long and has a dome of solid bone nearly nine inches thick. Illustration by John C. McLoughlin.

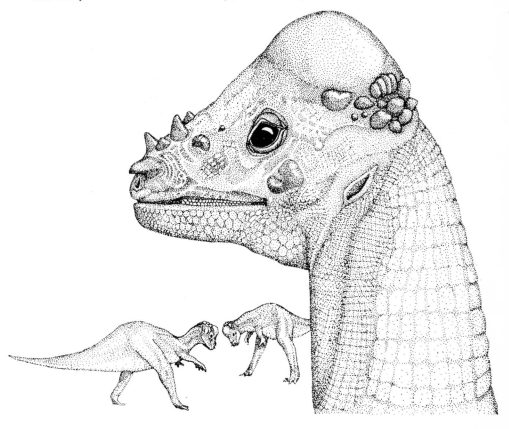

were covered with wartlike projections, was a land creature. A contemporary of *Tyrannosaurus,* it lacked the swimming ability of its hadrosaur cousins and probably often fell prey to the "Tyrant King."

The bone-heads relied perhaps too much on their grotesque ugliness and their thick skulls to ward off predators.

Horned Dinosaurs

Order—Ornithischia
Suborder—Ceratopsia
Family—Ceratopsidae
Genus—*Triceratops*
Monoclonius
Styracosaurus
Pentaceratops

Ceratopsians, or the horned dinosaurs, have always been paleontological favorites. The last of the reptilian giants, each species of ceratopsian dinosaur was characterized by a skeletal "frill" which extended from the back of the animal's skull, and covered its neck and shoulders. For more than a hundred years, scientists considered the frill protective. However, recent osteological studies of the ceratopsian's neck frill may alter the classic concept of its function, and subsequently, of the beast's appearance. A close examination of the frill's boney structure, and the physical relationship of muscle to bone in modern animals, have revealed another possibility.

Because of the tremendous pressures produced during a large animal's normal movement, muscle connections become strengthened with the aid of roughened, slightly depressed areas on the bone's surface. These irregularities, called muscle scars, are easily recognizable on fossil bones.

Paleontologists who are now studying these scars on ceratopsian frills believe, with regrets from an aesthetic point of view, that the frill itself was not a kind of protective armor, but rather a single, great connection for the muscles needed to support a head which may have weighed as much as two tons, and to operate the animal's lower jaw during feeding. If this is indeed the case, none of the frill would have protruded.

There is no doubt that a frill, even if buried under muscle, provided necessary protection for their leathery, but otherwise vulnerable necks. Many well-preserved frills exhibit deep, healed gouges, breaks, and teeth marks, suggesting that the horned dinosaur often fought its own kind—perhaps in territorial battles. And certainly the ceratopsians were not exempt from the predatory efforts of carnosaurs like *Tyrannosaurus*. Several species did grow long, slender spikes which would have protruded through the animal's skin as an added deterrent to the meat-eaters. *Styracosaurus,* an early form of ceratopsian had the grandest display of spines, and probably resembled a giant horned chameleon.

A large ceratopsian dinosaur had from one to three horns atop a skull which looked to be nine feet long. This was deceptive, however, for the skull vault may have been surrounded by the frill which would have accounted for up to six of those nine feet.

The fossil horns, though long and sharp, were only bony cores covered with a tough outer sheath of actual horn. The hooves and horns of modern cattle are constructed of this same material. The sheath added length and sharpness to the reptilian horn. When damaged, the horn regenerated the destroyed part, or if a large enough part was broken off, an entirely new horn was produced.

In multihorned species, like the *Triceratops,* one horn was set above each eye and one larger horn grew in the center of its face. The brain of *Triceratops,* the last of the horned dinosaurs, weighed close to two pounds, a dozen times more than other dinosaurs' brains. If not as "brainy" as a mammal, *Triceratops* at least may have had more highly developed instincts than most reptiles.

Fossil evidence suggests that the ceratopsians traveled in great herds, and may have formed protective circles around their young when attacked—a curious habit which the Musk Ox exhibits

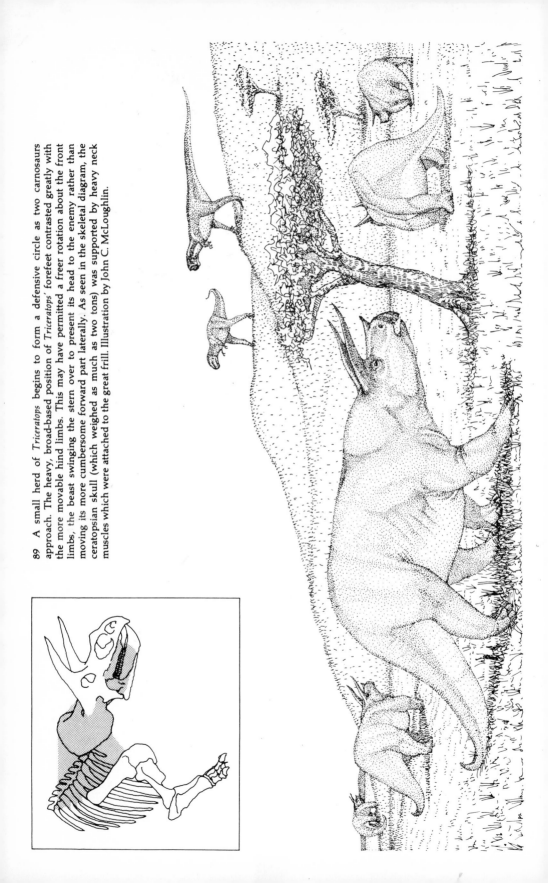

89 A small herd of *Triceratops* begins to form a defensive circle as two carnosaurs approach. The heavy, broad-based position of *Triceratops*' forefeet contrasted greatly with the more movable hind limbs. This may have permitted a freer rotation about the front limbs, the beast swinging the stern over to present its head to the enemy rather than moving its more cumbersome forward part laterally. As seen in the skeletal diagram, the ceratopsian skull (which weighed as much as two tons) was supported by heavy neck muscles which were attached to the great frill. Illustration by John C. McLoughlin.

90 *Barosaurus*, an Upper Cretaceous ceratopsian dinosaur. Notice the well-developed skull frill which extends well over the reptile's neck and back, acting both as protection and as support for its massive head. Photo courtesy Royal Ontario Museum.

91 Skeleton of *Triceratops*, an Upper Cretaceous ceratopsian dinosaur. From Marsh.

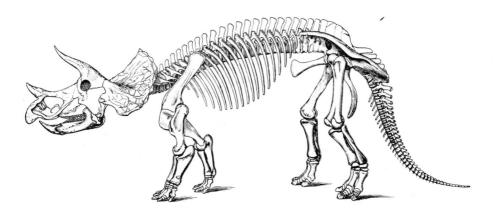

92 Skull of the horned dinosaur *Monoclonius*. Its long horn and beak were covered with a smooth horny material. About sixty-three inches in length. From Osburn.

93 *Chasmosaurus*, a horned dinosaur from the Upper Cretaceous. Photo courtesy Royal Ontario Museum.

today. If challenged, these great horned reptiles faced their opponents and perhaps charged at soft underbellies, inflicting often fatal puncture wounds.

Ceratopsians, the last of the reptilian giants, were perhaps the most successful of the dinosaurs. They were certainly the most abundant. One man observed the bones belonging to hundreds of *Triceratops* skeletons weathering from late Cretaceous rocks. Less than one of every hundred thousand dinosaurs ever became fossilized. Not including the uncounted millions of specimens lost

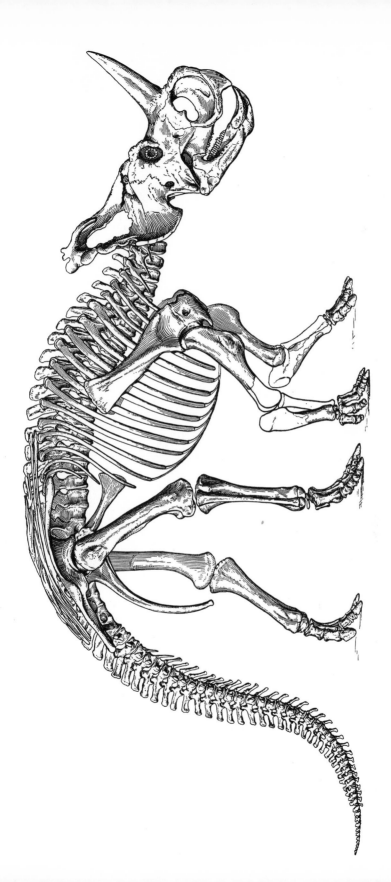

94   Skeleton of *Monoclonius*. Drawing by E. S. Christman.

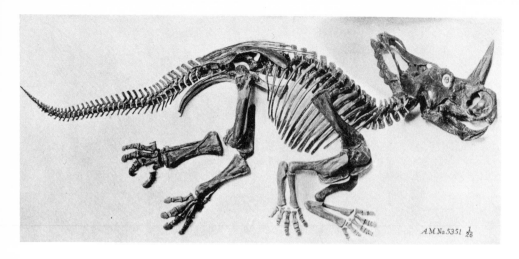

95   Skeleton of *Monoclonius*, as found. About eighteen feet in length. From Brown.

96   Early twentieth-century restoration of an Upper Cretaceous environment showing the eating habits of *Monoclonius*. Drawing by Richard Beckert.

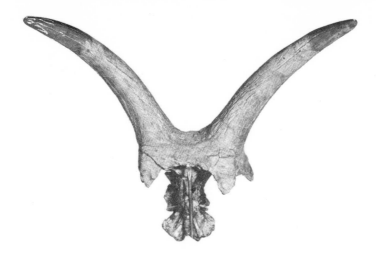

97   In 1887, Professor O. C. Marsh of Yale University described these horns as those of the buffalo *Bison alticornis*. Two years later, much to his embarrassment, the first ceratopsian dinosaur was discovered with this supposedly bison horns atop its great skull. Specimen found in the Denver beds on Green Mountain Creek, near Denver, Colorado. From Osburn.

to erosion since the end of the Mesozoic, those observed skeletons alone may represent 30 million or more living dinosaurs. Even these 30 million would be but a tiny percentage of the late Cretaceous *Triceratops* population. Nearly one hundred percent of the fossil dinosaurs still remain undiscovered, hidden under 70 million years of post-Mesozoic sediment accumulations.

Some evidence suggests that *Triceratops* may have existed in the Southwest into the earliest beginnings of the Cenozoic, or Age of Mammals, some 70 million years ago.

5

The End of an Era

By the end of the Cretaceous period the majority of the earth's ecologic characteristics had changed radically. The Mesozoic era came to a sudden, dramatic climax. Obviously there were reptile groups that persisted into the Tertiary era, but all forms of the dinosaur, with the possible exception of the *Triceratops,* were extinct. The end must have come suddenly for the "thunder lizards." Only their fascinating fossil record and a few distantly related modern reptiles tell us their story.

During this century, many theories have been developed to explain the nearly simultaneous extinction of the ruling reptiles some 70 million years ago. Some explanations have more merit than others: all must be highly speculative. Generally, extinctions are believed to result from the inability of a particular group of plants or animals to adapt to environmental change. Though the processes leading to extinction are quite simple, they are also irreversible. In the case of the dinosaurs, the slow geologic rise of the Rocky Mountain group began a chain of events—known as the Laramide revolution—which many reptile groups found intolerable. Low-lying flatlands once covered with swamps began to drain, and their life-giving waters disappeared. Desiccation of endless miles of swamps resulted in a rapid series of extinctions. As critical members of a long food chain disappeared, so did the dinosaurs. The aquatic reptiles, with their landlocked food chain, followed the dinosaurs into oblivion. So at least, goes one theory.

98 The rise of mammals at the close of the Mesozoic era may have contributed to the fall of the dinosaurs. This is an artist's conception of a multituberculate mammal, *Taeniolabis,* as it feeds on a dinosaur egg. Illustration by John C. McLoughlin.

But it must be remembered that we have a meager record of the untold generations of reptiles which succeeded each other through past ages. Each fossil discovery sheds some new light on the many unanswered questions surrounding this great extinction. The history of dinosaurs remains an unfinished one.

Following is a list of theories about the causes of the dinosaurs' extinction. It will be left to the reader to evaluate each. Of necessity, they reveal some of the inherent pitfalls of a new science.

1. It has been proposed, in rather romantic, Jules Verne fashion, that a great comet swept past the sun and engulfed the earth in its hot, poisonous tail during the Age of Reptiles, choking and cooking all of the dinosaurs.

2. The eggs of dinosaurs were all eaten by primitive mammals.

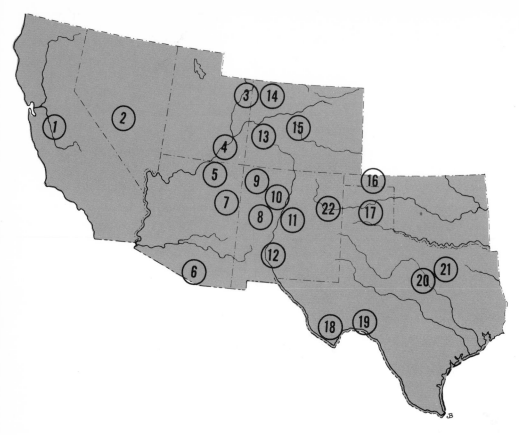

99 Dinosaur localities in the Southwest. Major areas where dinosaur fossils have been found.

1. San Joaquin Valley, California: Upper Cretaceous (hadrosaurs)
2. Ichthyosaur State Park, Nevada: Jurassic (marine reptiles)
3. Dinosaur National Monument, Jensen, Utah: Jurassic
4. San Juan County, Utah: Triassic-Cretaceous
5. Northern Arizona: Triassic-Cretaceous
6. Area locations south of Tucson, Arizona: Cretaceous
7. Petrified Forest National Park, Arizona: Triassic
8. Acoma Indian Reservation, New Mexico: Jurassic
9. Kirtland shale. Navajo Reservation, New Mexico: Upper Cretaceous
10. Ghost Ranch area, vicinity of Abiquiu, New Mexico: Triassic
11. Sandia Park area, New Mexico: Cretaceous
12. Elephant Butte, New Mexico: Cretaceous
13. Delta County, Colorado: Cretaceous
14. Moffat County, Colorado: Jurassic-Cretaceous
15. Denver Beds, Colorado: Jurassic-Cretaceous
16. Oklahoma panhandle: Cretaceous
17. Palo Duro Canyon, Texas: Triassic
18. Big Bend region, Texas: Lower Cretaceous
19. Brewster County, Texas: Cretaceous
20. Glen Rose, Texas: Lower Cretaceous
21. Dallas, Texas: Jurassic (marine reptiles)
22. Tucumcari, New Mexico: Triassic

3. Shells of the dinosaurs' eggs gradually became too thin, as birds' eggs today, to support the developing embryo within.

4. A great Ice Age at the close of the Mesozoic froze all of the dinosaurs.

5. The environmental temperature rose to such a level at the end of the Cretaceous that certain large reptiles were unable to survive. Many groups of smaller reptiles were able to survive in furrows during periods when the air's temperature rose above critical levels.

6. A great epidemic spread by insects conquered the giants.

7. Dinosaur populations became too large, setting off internal population controls: dinosaurs simply ceased to reproduce. This happens from time to time with modern animal groups, but, barring interference from people, the trend always reverses itself.

Though the Mesozoic giants were gone, a new class of animals was ready to take their place. The Age of Mammals began—a new story as diverse and fascinating as the one just finished. But let us not consider the extinct reptiles biological failures. Our own ancestry goes back less than four million years: the dinosaurs' reign lasted fifty times that.

Appendix

COLORADO

Denver Museum of Natural History, City Park, Denver.
> An excellent display of reconstructed dinosaur skeletons, as well as many other fascinating fossil and natural history displays.

University of Colorado Museum, Boulder.

Florissant Fossil Beds National Monument, Florissant.

NEW MEXICO

Ghost Ranch Museum, Abiquiu.
> One of the most charming little museums in the Southwest. The enclosed portion of the museum houses a very interesting display of the Triassic dinosaurs discovered on the ranch.

Geology Museum, University of New Mexico, Albuquerque.
> The one dinosaur specimen exhibited here, a large footprint of *Tyrannosaurus,* is worth seeing when in the area. The museum also exhibits many other fossils and minerals.

Natural History Museum, Carlsbad Caverns.

UTAH

Dinosaur National Monument, east of Vernal, situated in both Colorado and Utah.
> This is perhaps the most striking exhibit of dinosaur remains. Twenty-six nearly complete skeletons and innumerable isolated bones have been found here, and many are displayed in situ.

University of Utah Natural History Museum, Salt Lake City.
> The museum houses a fine collection of fossil specimens.

Bryce Canyon National Park, Bryce Canyon.

Brigham Young University, Provo.
> The museum contains both dinosaur and fossil remains from the Southwest.

Vernal Fieldhouse of Natural History, Vernal.
> Displays many kinds of Utah fossils.

Dallas Museum of Natural History, Dallas.
El Paso Centennial Museum, Texas Western College, El Paso.
Big Bend Historical Museum, Sul Ross State College, Alpine.
Panhandle-Plains Historical Museum, Canyon.
Department of Geology, Southern Methodist University, Dallas.
Texas Memorial Museum, University of Texas at Austin.
West Texas Museum, Texas Technological University, Lubbock.
University of Houston Geological Museum, Houston.
University of Texas Memorial Museum, Baird.

ARIZONA

Petrified Forest National Monument, Holbrook.
> An incredibly beautiful natural display of agatized trees weathering from the Chinle formation. Many fossil vertebrates are also found here.

Museum of Northern Arizona, Flagstaff.
> This museum is set in one of the most picturesque locations in America, and houses a good collection of vertebrate fossils.

Museum of Geology, University of Arizona, Tucson.
Tucson Mineral and Gem World, Tucson.
> A fascinating little museum of fossils, minerals, and other artifacts.

SOUTHERN CALIFORNIA

Los Angeles County Museum of History, Science and Art.
Occidental College, Los Angeles.
University of California at Los Angeles.
University of Southern California, Los Angeles.
Riverside Municipal Museum, Riverside.
San Diego Society of Natural History Museum and Library, San Diego.
Santa Barbara Museum of Natural History, Santa Barbara.
Museum of Paleontology, University of California, Berkeley.
Museum at the California Institute of Technology, Pasadena.
California Academy of Sciences, San Francisco.

OKLAHOMA

The Stovall Museum, University of Oklahoma, Norman.
Museum of the Great Plains, Lawton.

Glossary

acetabulum   cup-shaped pelvic depression into which the ball joint of the thigh bone fits.

Age of Fish   common name for the Paleozoic era. During this time, between 550 and 250 million years ago, the earliest of the reptiles began their long evolutionary history.

Age of Mammals   the last 70 million years of the earth's geologic history.

Age of Reptiles   an ecological pseudonym for the Mesozoic era. A 165-million-year period during which reptiles were the dominant form of vertebrate life.

alkali pond   small inland body of water containing soluble salts and hydroxides, chiefly the carbonates of sodium or potassium; usually poisonous to animal life.

ammonite (Ammon's Horn)   squidlike mollusk possessing a chambered shell separated into sections by sutures. The nearest living relative is the chambered nautilus.

amphibious   capable of living either on dry land or in water. Most sauropod dinosaurs were apparently amphibious.

Ankylosauria   suborder of ornithischian dinosaurs which developed a nearly complete covering of bony armor. Ten genera are represented in Late Cretaceous rocks of the Southwest.

Archeozoic era   the era in which most of the earth's landforms and seas developed. Although simple one-celled life first developed in the Late Archeozoic, the rocks of this age have revealed few, if any, fossils. Archeozoic rocks are thought to be more than two billion years old.

archosaurian   "ruling reptiles"; a biological subclass which includes dinosaurs, crocodiles, and flying reptiles.

Cambrian   the first period of the Paleozoic era, beginning about 520 million years ago.

carnivorous   meat-eating.

106

Carnosauria   large, flesh-eating dinosaurs, like *Tyrannosaurus rex.*

Cenozoic era   the fourth major era of geologic time commonly called the Age of Mammals; preceded by the Mesozoic era and followed by the Holocene period.

Cephalopoda   class of shelled, marine invertebrates common during the Age of Reptiles. Living representatives are the squid, octopus, and nautilus.

ceratopsians   horned dinosaurs, like the *Triceratops.*

champsosaurs   abundant Late Cretaceous needle-toothed pseudocrocodiles.

coelurosaurs   small theropod (carnivorous) saurischian dinosaurs.

coprolite   fossil dung.

Cretaceous   the last of the Mesozoic era's periods, beginning about 120 million years ago and ending with the start of the Tertiary period some 70 million years ago.

cross-bedding   layers of sediment which are not deposited parallel because of changes in wind direction or water flow.

crossopterygian lobe-fin   fish ancestral to the amphibians. Nostrils or skull roof may have been used in conjuction with lungs rather than with gills.

cycads   order of flowering gymnosperms which became the dominant Mesozoic vegetation from the Triassic to the Early Cretaceous.

dermal armor   small plates of bones which grew in the flesh of some dinosaur species, acting as protection for an otherwise vulnerable skin. These platelets were not connected in any way to the animal's skeleton.

Devonian   the fourth period of the Paleozoic era, between 345 and 405 million years ago.

endothermic   cold-blooded. Like most reptiles, animals that are incapable of maintaining a constant body temperature.

epoch   subdivision of geologic periods, such as the Paleocene, Eocene, Oligocene, Miocene, Pliocene, and Pleistocene epochs of the Tertiary period.

eras   the major divisions of geologic time, like the Paleozoic, Mesozoic, and Cenozoic.

evolution   changes in organisms caused by slow alterations of their genetic makeup.

extinction   the total disappearance of a species, often the death of one animal.

fossil   any evidence of life from past geologic ages. Bones, impressions of invertebrates, and footprints are examples.

gastroliths   stomach stones. Highly polished rocks found in the stomach areas of herbivorous dinosaurs' skeletons.

geologic column   the simple relationship of older to younger rocks in their relative order of deposition, with the older sediments usually being lowest on the column.

geologic time scale   the age of the earth, usually divided into segments of time set apart by significant evolutionary events.

gymnosperms   plants whose seeds are unprotected by any covering. The name derives from the Greek *gymnos* (naked) and *sperma* (seed). Modern examples include shrubs, evergreens, and cycads.

homothermic   warm-blooded. Capable of maintaining a constant body temperature despite external environmental conditions.

Ice Age   common name for the Pleistocene, which in fact experienced several distinct ice ages.

instincts   an animal's inborn, unlearned responses to environmental stimuli which drive an animal to behave in a way characteristic to a species.

Jurassic period   the second period of the Mesozoic era, beginning 150 million years ago and lasting for 25 million years.

Laramide revolution   the gradual geologic uplift which formed the Rocky Mountains near the end of the Cretaceous period.

Mesozoic era   the third major era of geologic time, commonly called the Age of Reptiles. Preceded by the Paleozoic and followed by the Cenozoic.

metagenesis   the development or alteration of a generation in which the descendants develop into forms quite unlike the parent.

Mississippian   the fifth period of the Paleozoic era, between 345 and 310 million years ago.

mitosis   the biological process by which a cell reproduces itself through division of its plasma and genetic material.

Morrison Formation   the best known dinosaur-bearing strata of the Upper Jurassic period. The Morrison, dark gray-green shale with local beds of sandstone, is exposed in many states.

Multituberculata   order of primitive Mesozoic mammals ranging in size from that of a mouse to that of a beaver.

Ordovician   the second period of the Paleozoic era, between 425 and 500 million years ago.

Ornithischia   order of dinosaurs which developed birdlike pelvic structures.

orogeny   the process of extensive mountain systems uplifting due to geologic pressure over long periods of time.

ossicle   bony subskin platelets which act as secondary armor in many dinosaurs.

paleogeography   the geography of ancient landforms.

paleontology   the study of fossil life. Divisions of the science are vertebrate, invertebrate, and paleobotany.

Paleozoic era   the first era of geologic time in which distinct forms of complex life developed.

Pennsylvanian   sixth period of the Paleozoic era, commonly called the Age of Coal.

periods   major time-life divisions within the geological eras.

Permian   the seventh period of the Paleozoic era, between 230 and 280 million years ago.

phytosaurs   semiaquatic, crocodilelike thecodonts common during the Triassic period.

Pleistocene   geologic period, commonly called the Ice Age, from 2 million to 40 thousand years ago. It was the age of giant mammals; dinosaurs had already been extinct for nearly 70 million years.

Precambrian   the 2,300,000,000-year period, for the most part void of fossil evidence of life.

Proterozoic era   era preceding the Paleozoic in which complex life-forms evolved.

Pterosauria   order of archosaurian flying reptiles.

Quaternary   geologic period comprising the last one and a half million years of the earth's history.

Reptilia   class of vertebrates which includes snakes, lizards, turtles, alligators and crocodiles, dinosaurs, mosasaurs, plesiosaurs, and a variety of other extinct life-forms. The reptiles evolved from amphibians and gave rise to birds and mammals.

Saurischia   order of dinosaurs which developed reptilelike pelvic structures.

sauropods   suborder of saurischian dinosaurs, some of which became the largest nonmarine animals ever to have lived.

Silurian   the third period of the Paleozoic era, from 405 to 425 million years ago.

Stegosauria   suborder of armored, ornithischian dinosaurs which were abundant during the Jurassic period.

submergence   a downwarping of large land surfaces and their subsequent coverage by seas or oceans.

taxonomy   science of naming and classifying biological organisms.

Tertiary period   geological period which lasted 63 million years and is considered the Age of Mammals.

thecodont   Triassic reptile with teeth in jaw sockets.

theropod   characteristic reptilian carnivores of the Late Triassic, Jurassic, and Cretaceous periods.

Triassic period   first geologic period in the Mesozoic era, beginning some 200 million years ago and ending 35 million years later.

trochanter   flange of bone to which strong locomotor muscles are connected.

vertebrates   fish, amphibia, birds, mammals, and reptiles, the five biological classes which belong to the phylum chordata and possess a nerve cord surrounded by a bony vertebral column.

volcanism   eruption of lava and gas from beneath the earth's surface. Resultant poisonous gases may have contributed to the extinction of some dinosaur forms.

Bibliography

The following books are recommended to those interested in learning more about dinosaur paleontology of the Southwest, as well as about the evolution of vertebrate life in general. They range in scope from lay and juvenile to technical works by noted paleontologists. The list is far from inclusive. Further information concerning specific areas can be obtained from state or local United States Geological Survey offices, a list of which follows the bibliography.

Andrews, R. C. *All About Dinosaurs.* New York: Random House, Inc., 1953.

Brown, B. "The Cretaceous Ojo Alamo Beds of New Mexico with Descriptions of the New Dinosaur Genus *Kritosaurus."* *Bull. of the American Museum of Natural History* 28 (1910):267–74.

Brown, B. and Schlaikjer, E. "The Rise and Fall of the Dinosaurs." *Natural History* 48 (1941).

Camp, Charles L. *A New Type of Small Bipedal Dinosaur from the Navajo Sandstone of Arizona.* Berkeley: University of California Press, 1936.

Colbert, E. H. *The Dinosaur Book; Ruling Reptiles and Their Relatives.* New York: American Museum of Natural History Press, 1945.

————. *The Dinosaur World.* New York: Stravon Educational Press, 1976.

————. *Evolution of the Vertebrates.* New York: John Wiley and Sons, 1965.

————. *Men and Dinosaurs.* New York: E. P. Dutton & Co., Inc., 1968.

————. *The Age of Reptiles.* New York: W. W. Norton & Company, Inc., 1965.

Desmond, Adrian. *The Hot-Blooded Dinosaurs.* New York: Dial Press, 1976.

Dunbar, Carl O. and Waage, Karl M. *Historical Geology.* 3d ed. New York: John Wiley & Sons, Inc., 1969.

Gilmore, C. W. "Osteology of the Jurassic reptile *Camptosaurus,* with a revision of the species and genus, and description of two new species." *Proc. U. S. Nat. Mus.* 36 (1909):197–332.

———. "Osteology of the Carnivorous Dinosauria in the United States National Museum, with special reference to the Genera *Antrodemus (Allosaurus)* and *Ceratosaurus."* *Bull. U. S. Nat. Mus.* 110 (1920):1–159.

———. "Osteology of the Armored Dinosauria in the United States National Museum, with special reference to the Genus *Stegosaurus."* *Bull. U. S. Nat. Mus.* 89 (1914):1–143.

———. "The Horned Dinosaurs." *Smithsonian Annual Report* (1920).

———. "A New Sauropod Dinosaur from the Ojo Alamo Formation of New Mexico." *Smithsonian Annual Report* (1922).

Good, John, et al. *The Dinosaur Quarry.* Washington, D.C.: National Park Service, 1958.

Moore, R. C. *Introduction to Historical Geology.* New York: McGraw-Hill Book Co., 1958.

Moore, Ruth. *Man, Time, and Fossils.* New York: Alfred A. Knopf, Inc., 1953.

Pangborn, M. W., Jr. *Earth for the Layman.* Washington, D.C.: American Geological Institute, 1957.

Raymond, P. E. *Prehistoric Life.* Cambridge, Mass.: Harvard University Press, 1950.

Richards, H. G. *Record of the Rocks.* New York: Ronald Press, 1953.

Romer, A. S. *Vertebrate Paleontology.* Chicago: University of Chicago Press, 1945.

———. *Osteology of the Reptiles.* Chicago: University of Chicago Press, 1956.

Simpson, G. G. *Life of the Past.* New Haven: Yale University Press, 1953.

Stirton, R. A. *Time, Life, and Man: The Fossil Record.* New York: John Wiley and Sons, 1959.

White, Theodore E. *Dinosaurs at Home.* New York: Vantage Press, 1967.

STATE GEOLOGICAL SURVEYS

Arizona
College of Mines
Arizona Bureau of Mines
University of Arizona
Tucson, Arizona 87525

California
Division of Mines and Geology
Department of Conservation
Ferry Building
San Francisco, California 94111

Colorado
Colorado Metal Mining Fund Board
204 State Office Building
Denver, Colorado 80202

Nevada
Nevada Bureau of Mines
University of Nevada
Reno, Nevada 89507

New Mexico
State Bureau of Mines and Mineral Resources
New Mexico Institute of Mining and Technology
Socorro, New Mexico 87801

Oklahoma
Oklahoma Geological Survey
University of Oklahoma
Norman, Oklahoma 73069

Texas
Bureau of Economic Geology
University of Texas
Austin, Texas 78712

Utah
Utah Geological and Mineralogical Survey
103 Civil Engineering Building
University of Utah
Salt Lake City, Utah 84102

Index

Age of Fish. *See* Paleozoic era
Age of Mammals. *See* Cenozoic era
Age of Reptiles, 9–14 passim. *See also*
    Mesozoic era
Alamosaurus, 41
alkali pond, 28
American Museum of Natural History
    (New York), ix, x, 25, 87
Anchisaurus, 30
Ankylosaurus, 79–81
Antrodemus (*Allosaurus*), vii, 34n, 55–60
    passim, 72
Apatosaurus, 32, 34n, 41, 43, 45
Arizona, ix

Baldwin, David, discovers *Coelophysis*, 25
birds, 5, 14, 18–19, 70
Brachiosaurus, 19, 34n, 41
Brontosaurus. *See Apatosaurus*
Buckland, William, 19

California, ix
Camarasaurus, 34n, 41, 45
cannibalism, 25
Carnegie Museum (Philadelphia), 34
carnivores, 16, 32, 40, 43, 48, 49, 55–60
    passim, 64, 65, 66, 71–75, 79, 82, 93
Cenozoic era, 24, 99
cephalopods, 14
ceratopsians, 64, 92–94
Ceratosaurus, 34n, 60–64
champosaurs, 70
Coelophysis, vii, 25–29, 48
Coelurus (*Ornitholestes*), 34n, 48–49
cold-bloodedness. *See* endothermy
Colorado, ix, 24–29 passim, 32–34, 65, 75
Corythosaurus, 82
Cretaceous period, vii, 8, 9, 65–99 passim,
    100
cross-bedding, 40–41
cycads, 9. *See also* plants

deinodonts, 60, 72, 75
Didelphis, 70
Dinosaur National Monument, x, 34–35, 40
Diplodocus, 34n, 41, 45

endothermy, 18–19

epoch, defined, 8
era, defined, 8
evolution, 2, 5, 8, 16, 30
extinction of dinosaurs, 71, 100–103

gastroliths, 45
geologic column, 8
geologic time scale, 8
glyptodonts, 79
Gorgosaurus (*Albertosaurus*), 54, 75

hadrosaurs, 81–82, 87
herbivores, 16, 43, 48, 65, 66, 79
Hesperornis, 70
homothermy, 18–19

Ice Age, 4, 79. *See also* Pleistocene
    period
Icthyornis, 70
International Committee of Nomenclature,
    7

Jurassic period, 8, 9, 30, 32–64, 70

Kritosaurus, 82

Lambeosaurus, 82
Laramide revolution, 100
Linnaeus. *See* von Linné, Karl

mammals, vii, 14, 16, 18–19, 29, 34n, 45, 70,
    93
marsupials, 14
Mesozoic era, 8, 9–14 passim, 20, 24–99,
    100, 103
Monoclonius, 92
Morrison Formation, 32–34, 40–41
mosasaurs, 70
multituberculates, 14

National Park Service, x, 35
Nevada, ix
New Mexico, ix, 14, 24–29 passim, 65, 75

Oklahoma, ix
Ornitholestes. *See Coelurus*
Ornithomimus, 75–77
Owen, Sir Richard, 14

Pachycephalosaurus, 91
paleontology, 4, 5–8 passim
Paleozoic era, 20, 24
Parasaurolophus, 82
Pennsylvanian period, 20
Pentaceratops, 92
period, defined, 8
Permian period, 8
Petrified Forest (Arizona), 29
Phobosuchus, 70
phytosaurs, 29
Placerias, 29
plants, as food source, 9, 14, 29. See also
   herbivores
Plateosaurus, 30
plesiosaur, 70
Precambrian period, 8
Pteranodon, vii
pterodactyle, 16
pterosaurs, 9

regeneration, 45
Rutiodon, 29

sauropods, 19, 30, 41–45, 70–71
sclera, 22
Smithsonian Institution, ix, x

Stegosaurus, 32, 34n, 40, 49–55 passim, 79
Struthiomimus, 75–77
Styracosaurus, 92, 93
Systema naturae (von Linné), 6

taxonomy, 5–9
Taylor, Bert Leston ("The Dinosaur"),
   54–55
Tertiary era, 100
Texas, ix, 29, 70, 75
thecodonts, 16, 20, 22, 30
Trachodon (Anatosaurus), 81–82
Triassic period, 8, 9, 24–31, 48
Triceratops, 92, 93–99 passim, 100
Tröodon, 91
Tyrannosaurus rex, 16, 71–75, 92, 93

United States Geological Survey, ix
Utah, ix, 24–29 passim, 32, 75

vertebrates, 2, 41
volcanism, 14
von Linné, Karl, 5–6

warm-bloodedness. See homothermy
Whitaker, George, 25–28